'O' Grade Chemistry

Essential facts and theory

Hilary Cockburn

'O' Grade Chemistry

Essential facts and theory

R. A. Robertson

Edward Arnold

First published 1975
by Edward Arnold (Publishers) Ltd 25 Hill Street, London W1X 8LL

ISBN 0 7131 1942 X

Acknowledgements.

The publishers wish to thank Blackie and Son Ltd., for permission to reproduce the tables on p 34, 35, 36, 37, 39, 40 and 41 of *Three Figure Mathematical Tables and Science Data.* These tables appear as an appendix at the end of this book.

Text set in 11/12 pt. IBM Press Roman, printed by photolithography, and bound in Great Britain at The Pitman Press, Bath.

Preface

As long as the examination system exists, there will be a need for a school textbook which covers the 'O' Grade syllabus as briefly as possible, and in a manner suitable for revision purposes. In addition, a textbook of this sort is of great value as a back-up to class work.

This book is not intended to be used as a textbook-workbook; in general, it gives the essential theory and the results which would be obtained from experiments without elaborating on the experimental details. Not only does this condense the book for revision purposes, but it also leaves the class teacher free to devise his own worksheets if he so wishes, and to modify or add experiments in such a way as to reinforce the essential theory covered in the book.

Since a set of data tables is supplied in the 'O' Grade Examination, pupils should be encouraged to use these tables at all times to give them maximum help when writing formulae, ion electron equations, carrying out calculations or deducing information from the Periodic Table.

Often a pupil finds difficulty when writing formulae, because the approach has frequently been to teach the formation of the ionic or covalent bond and, using this information, to write formulae. The pupil was encouraged to decide whether a compound is ionic or covalent and then, obtaining ionic charges or numbers of bonds from the Periodic Table, to write the formula. The process of learning to write formulae can be simplified if the pupil does not have to worry about which type of bond is present. The problem of ionic or covalent character need only be considered when necessary, and certainly after the original 'covalent type' formula has been written. Writing the formula by this method necessitates the pupil obtaining a 'combining number' or 'valency' from the data tables (as indeed, he does anyway when he uses the ionic charge or the number of covalent bonds to write a formula). The use of the word valency went out of fashion in school chemistry because of its connection with the rote learning of the traditional syllabus, but since it is a widely accepted term I have used it in this book.

At the end of each chapter are a number of revision questions which the reader is asked to answer; he may then check his answer from the text. These questions are intended purely for revision purposes, and are not in any way intended to be 'O' Grade type questions.

Although I have a personal preference to teach the work in a somewhat different order from that suggested in the syllabus, I have adhered mainly to the syllabus order, leaving the teacher free to teach the work in the order he prefers.

I would very much like to thank Mr. W. Ross, Advisor in Science in Lanarkshire. Many of his ideas are incorporated in the text. I would also like to thank Mr. W. J. Clelland, Principal Teacher of Chemistry at Bellshill Academy, for his constructive criticism of part of the manuscript, and the Chemistry Staff at my school, Mr. R. Burgess and Mrs. N. Nimmo, for their assistance in testing the content of the book. My thanks must also go to the very helpful secretarial staff at my school who typed the original class notes which formed the basis of this book.

Contents

1

Atoms and atomic structure

1.1 Particles

All materials are made up of particles which are continuously in motion and, due to this moving energy (Kinetic Energy), have spaces between them. Some experiments showing this are given below.

1 Diffusion
Liquids and gases gradually mix, regardless of density.

Fig. 1 Bromine gas and air will mix

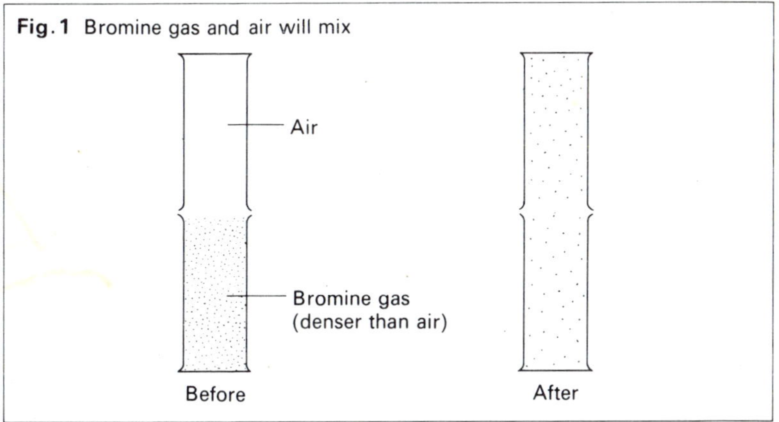

2 Experiments on mixing
As shown in Fig. 2:
50 cm^3 alcohol + 50 cm^3 water $\longrightarrow$ 98 cm^3 mixture.

There must, therefore, be spaces between the particles.

3 Smoke – Brownian Movement
Carbon 'specks' in smoke can be seen in the microscope to be jerking about in all directions – they are being bombarded by fast moving but invisible air particles.

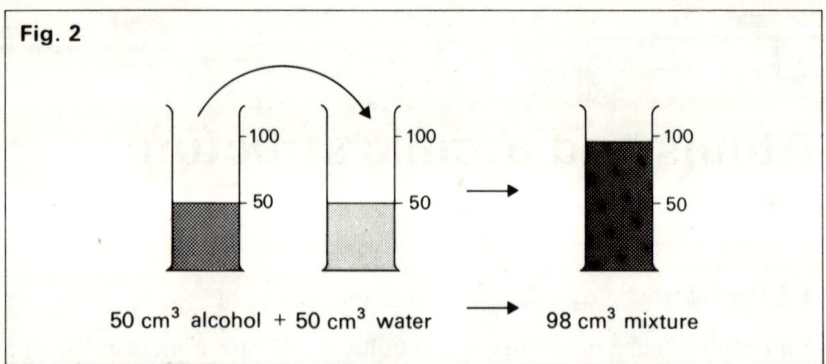

1.2 Atoms

All *compounds* are composed of atoms of one or more *elements* grouped together. There are just over 100 known elements, the first 92 of them are found naturally and the remainder made artificially.

The elements are composed of tiny particles called *atoms,* and atoms of different elements differ in size and mass.

1.3 Structure of the Atom

Before 1900, the atom had been thought to be a small, hard ball. However, experiments indicated the existence of a number of particles: the proton, the neutron, the electron, and many others.

The electron

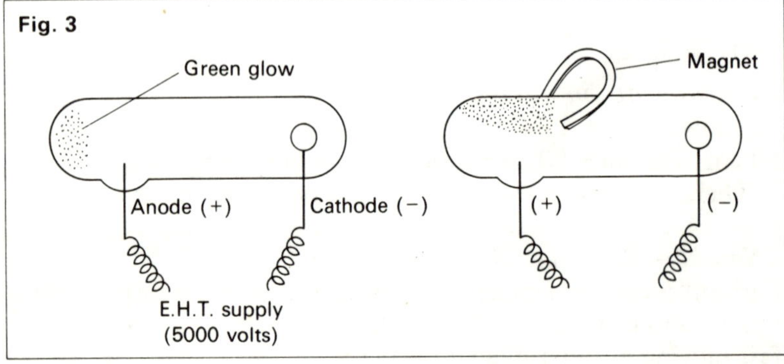

A cathode-ray tube shows a green, glowing region which can be bent by a magnet, indicating charged particles streaming away from the negative cathode (Fig. 3).
This shows the existence of negatively charged particles, called *electrons.*

The positive nucleus

In 1909, alpha particles were fired at very thin gold foil (Fig. 4).

Unexpectedly, most of the alpha particles passed straight through and only a few were deflected (about 1 in 10 000). In 1911, Rutherford suggested that most of the mass of the atom was concentrated in a tiny, positive *nucleus* at the centre of the atom, surrounded by negatively charged electrons which take up most of the space.

The positively charged particles in the nucleus were called *protons.*

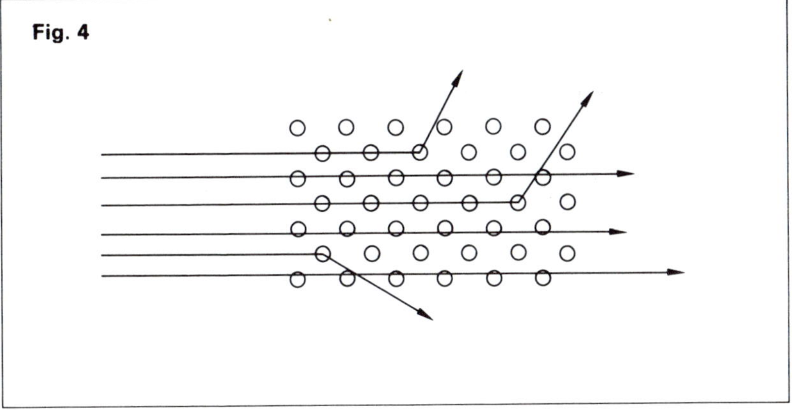
Fig. 4

1.4 Atomic mass scale

0. 000 000 000 000 000 000 000 000 001 67 kg is the mass of a proton. This is obviously too difficult to work with, so a new scale of mass is used – *the atomic mass scale.*

On this scale, the mass of a proton = 1 amu (atomic mass unit) and the mass of an electron = about $\frac{1}{2000}$ amu.

Atomic Structure

Three main particles have been shown to be present:

		mass	charge
nucleus	proton	1 amu	+1
	neutron	1 amu	0
	electron	$\frac{1}{2000}$ amu	−1

Arrangement of atoms in the periodic table

The elements are arranged in order of the number of protons which the atoms have. Since the number of protons does not change (the nucleus being protected in the centre of the atom), this number is called the *Atomic Number,* and tells us to which element an atom belongs.

The Atomic Number of an element is the number of protons in the nucleus of one of its atoms.

For example, all atoms of sodium, Na (atomic number = 11) contain 11 protons; all atoms of uranium, U (atomic number = 92) contain 92 protons.

Now, atoms of an element are electrically neutral, so the number of protons must equal the number of electrons. That is: Number of protons = Number of electrons.

For example, sodium atoms (atomic No. = 11) contain 11 protons and 11 electrons.

Since the electrons have little or no mass, it is the protons and neutrons in the nucleus which make up the mass of an atom and the total number of protons and neutrons is called the *Mass Number.*

The Mass Number of an atom is the number of protons plus the number of neutrons in its nucleus.

1.5 The Mass Spectrometer – Determining Mass Number

If a neutral atom has an electron pulled off it, it leaves the atom with a positive charge. A charged atom is called an *ion.* If these positive ions are passed through an electric or magnetic field, they are deflected, the lighter particles being deflected most (Fig. 5).

Particles of different mass strike the detector at different points, and a chart is obtained which shows the mass and proportion of each ion present (Fig. 6).

For example, chlorine gas shows atoms of two different masses to be present.

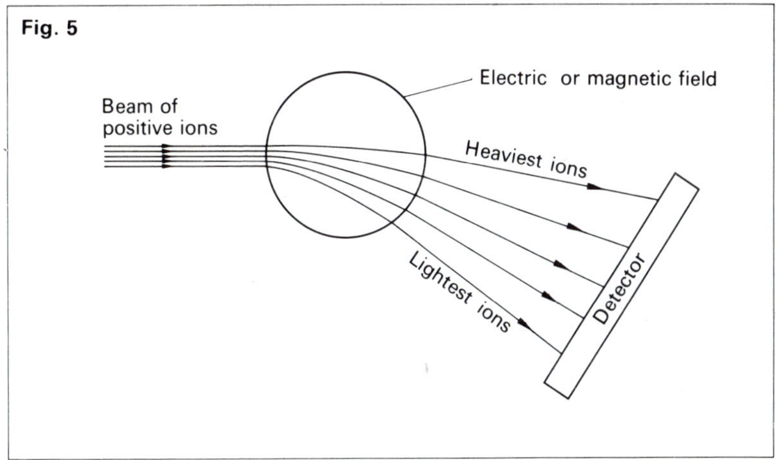

Fig. 5

The position of the lines tells us the mass of the atoms, and the height tells us the proportion present in the sample. Thus the mass spectrometer shows that, although atoms of one element must all have the same number of protons (the same Atomic Number), *they can contain different numbers of neutrons.*

For chlorine, the atomic number is 17. Therefore all chlorine atoms must have 17 protons. This means that the chlorine atom of mass number 35 must have 17 protons + 18 neutrons. The

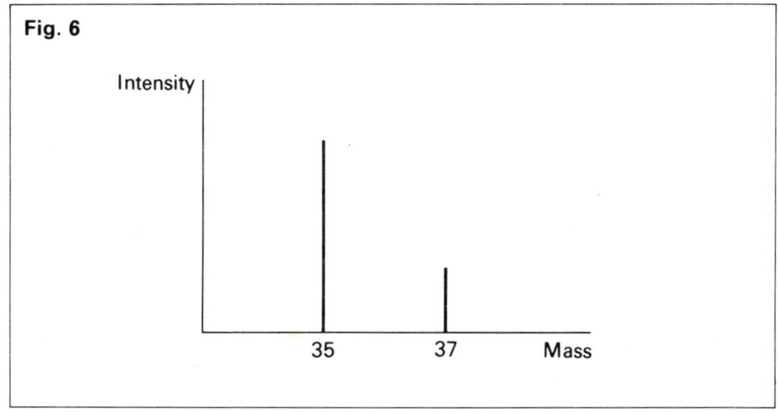

Fig. 6

chlorine atom of mass number 37 must have 17 protons + 20 neutrons.

These different chlorine atoms are called *isotopes* of chlorine. **Isotopes are atoms of an element with different mass numbers.** In other words, isotopes have the same number of protons but a different number of neutrons.

1.6 Representing atoms

We use a symbol such as $^{23}_{11}Na$ to represent an atom of sodium.

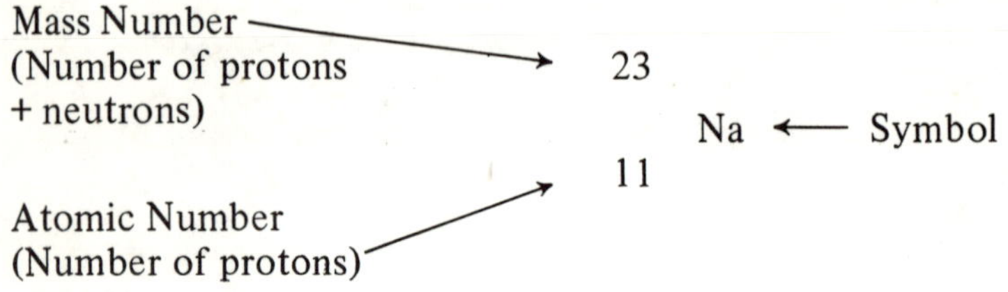

Atomic Mass (Atomic Weight)

Most elements contain a mixture of isotopes, so it is necessary to know the average mass of atoms of an element. This is the Atomic Mass.

The Atomic Mass of an element is the average mass of an atom of the element, taking into account the proportion of each isotope present.

For example, the element *chlorine* has two isotopes, $^{35}_{17}Cl$ (75%) and $^{37}_{17}Cl$ (25%). Since there are more $^{35}_{17}Cl$, the average mass works out at 35.5 amu. That is, the Atomic Mass of Chlorine = 35.5 amu.

1.7 The electrons

The negatively charged electrons are found outside the nucleus. They have energy which prevents them being pulled into the nucleus by the positively charged protons.

The electrons occupy various *energy levels* outside the nucleus, and since they are on the outside, they are responsible for the behaviour of an atom in a chemical reaction.

We can represent the electron structure of atoms as in the chart below. Carbon, for example, has two electrons in the first energy level and four in the next.

Electron structure of elements 1–20

H 1							He 2
Li 2.1	Be 2.2	B 2.3	C 2.4	N 2.5	O 2.6	F 2.7	Ne 2.8
Na 2.8.1	Mg 2.8.2	Al 2.8.3	Si 2.8.4	P 2.8.5	S 2.8.6	Cl 2.8.7	Ar 2.8.8
K 2.8.8.1	Ca 2.8.8.2						

Elements in the same column have the same number of outer electrons. They behave in a similar manner. The elements in the column with filled energy levels (helium, neon, argon, etc.) are very stable, and are called the *Noble Gases.*

The first energy level is filled with 2 electrons. The second energy level is filled with 8 electrons. The other energy levels are also filled with 8 electrons, but at a later stage this number can be increased.

1.8 Information, and where to find it

The information giving the *Atomic Number, Atomic Mass,* and *Electron Structure* of the elements can be obtained in the Periodic Table at the end of this book.

Some revision questions

You should try to answer each question and then check your answer by referring to the section indicated in brackets after the question.

1 Copy and complete the following table, listing the three main particles in the atom. (Section 1.4).

Particle	Mass (a.m.u.)	Charge
proton	1 a.m.u.	+1
neutron	1a.m.u	0
electron	$\frac{1}{1850}$ a.m.u	–1

2 What is meant by: (i) atomic number (ii) mass number (iii) isotopes (iv) atomic mass (atomic weight)? (Section 1.4 to 1.6).
3 Carefully redraw this table and fill in the blanks.

Symbol	Atomic Number	Mass Number	No. of protons	No. of neutrons	No. of electrons
$^{23}_{11}Na$	11	23	11	12	11
$^{16}_{8}O$	8	16	8	8	8
$^{35}_{17}Cl$	17	35	17	18	17
$^{81}_{35}Br$	46	81	35	46	35
$^{32}_{16}S$	16	32	16	16	16
$^{24}_{12}Mg$	12	24	12	12	12

(Section 1.4 to 1.6).
4 Give the electron arrangement for the atoms of atomic number (i) 15, (ii) 7, (iii) 18, (iv) 3, (v) 20. (Section 1.7.)

2.
(i) Atomic number = no. of protons or electrons in the element
(ii) Mass number = no. of protons + no of neutrons in the element.
(iii) Isotopes of an element have the same Atomic number, but different mass number
(iv) Atomic weight of an element is the weight of one atom of it taken on a scale were the atomic mass of carbon is 12.

4.
(i) 2.8.5 (ii) 2.5 (iii) 2.8.8 (iv) 2.1 (v) 2.8.8.2

2

How atoms combine

We can get some idea of how atoms combine if we look at the electrical properties of materials.

2.1 Conductors and non-conductors

Experiments showing the ability of materials to conduct electricity indicate three different types.

(a) *Metallic conductors* – good conductors of electricity in the solid or liquid state. They are not decomposed (broken up) by electricity.

(b) *Electrolytes* – substances which conduct electricity when dissolved in water, or when molten. These substances are decomposed by electricity. Electrolytes are usually solutions of acids, alkalis or salts, or molten salts.

(c) *Non-conductors* – do not appear to conduct electricity whether solid, liquid or in solution. Non-conductors are usually non-metallic elements, or compounds containing non-metallic elements.

Of these three we must now look at *electrolytes* and *non-conductors* in more detail.

2.2 Electrolytes

Let us look at the sort of results we get from electrolysing salts.

	Results of electrolysis	
Electrolyte	**At cathode (negative electrode)**	**At anode (positive electrode)**
Copper chloride solution	Copper	Chlorine
Molten lead bromide	Lead	Bromine

Since the copper and lead are always discharged at the cathode (the negative electrode) then we can say that the copper in the copper chloride and the lead in the lead bromide must have a positive electric charge (opposite charges attract) and be free to move to the negative electrode.

In the same way the chlorine in the copper chloride and the bromine in the lead bromide must have a negative charge and be free to move to the positive electrode.

Fig. 7 Electrolysis of (a) a solution of a salt, (b) a molten salt.

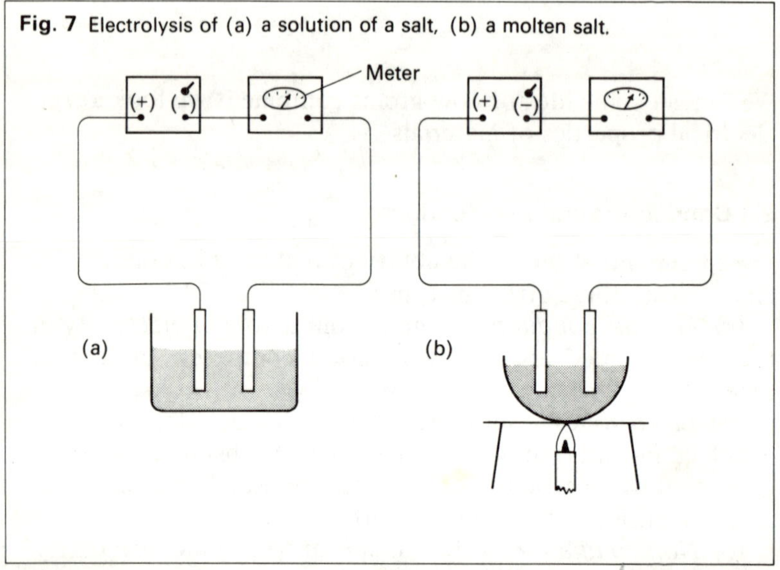

We call these charged atoms *ions.*

Since salts conduct electricity in solution or when molten, but not when in the solid state, ions must be held in position in the solid state, and become mobile only when the substance is dissolved in water or melted.

2.3 Non-conductors

These substances do not conduct whether in the solid state (e.g. sugar), the liquid state (e.g. alcohol or molten wax), or in solution (e.g. sugar solution).

This suggests that the particles in non-conductors must be uncharged. We call these uncharged particles *molecules.*

2.4 Bond formation and electrons

For atoms to combine they must collide, and on collision it is their outer electrons which must meet.

We have already shown the electron structure of some atoms. For example, the third row of the Periodic Table consists of:

Na	Mg	Al	Si	P	S	Cl	Ar
2.8.1	2.8.2	2.8.3	2.8.4	2.8.5	2.8.6	2.8.7	2.8.8

Within each energy level, the electrons are found in compartments called *orbitals* (or electron clouds), each orbital (or cloud) being able to hold 2 electrons (an electron pair).

What is an orbital?

An orbital (or electron cloud) is a space occupied by 1 or 2 electrons. Calculations show that the orbitals seem to have a particular shape, and the electrons are moving about so rapidly within these shapes that the orbital can be thought of as a negatively charged cloud.

Orbitals and the energy levels

The *first energy level* has 1 orbital and can hold 2 electrons. The *other energy levels* have 4 orbitals and can hold 8 electrons.

(The third level and those further away from the nucleus can at later stages hold more than 8 electrons, but this can be ignored at present.)

Since it is the electrons in the outer energy levels which are involved in collisions between atoms, we need only consider the outer levels.

When the atoms are combining, the orbitals in the outer energy levels are arranged as far apart from each other as possible, and take up a 3-dimensional tetrahedral shape.

Fig. 8 Tetrahedral arrangement of orbitals

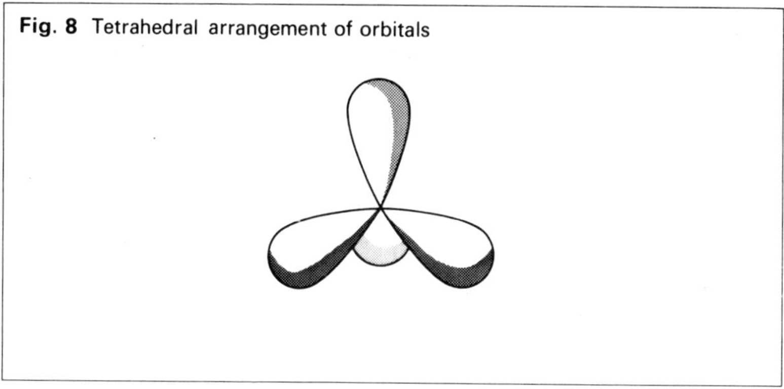

The electrons, if it is possible, occupy separate orbitals in the energy level, so no two electrons occupy the same orbital if there are others vacant.

The outer electron structure of the first 20 atoms can therefore be represented as in the chart below.

First Energy Level	H·	He∣
	half-filled orbital	filled orbital (electron pair)

Other levels:

Group	1	2	3	4	5	6	7	8
Second Level	Li·	Be·	·B·	·C·	·N·	∣O·	∣F·	∣Ne∣
Third Level	Na·	Mg·	·Al·	·Si·	·P·	∣S·	∣Cl·	∣Ar∣
Fourth Level	K·	Ca·	etc.					

From the table, the number of half-filled orbitals, filled orbitals or empty orbitals in any particular atom can readily be seen.

For example,

Na·	1 half-filled orbital, 3 empty orbitals.
·C·	4 half-filled orbitals.
∣Cl·	1 half-filled orbital.
H·	1 half-filled orbital (no empty orbitals, since the first energy level only has 1 orbital).

The noble gases (helium, neon, argon, etc.) are very stable. When atoms combine, they tend to achieve the electron structure of the noble gases which is a full outer energy level (all the orbitals filled).

2.5 How Atoms Combine

All atoms combine in the same way. Before they can join together, they must *collide,* and their half-filled orbitals overlap. (That is, their half-filled electron clouds merge.) Fig. 9 shows this happening to two atoms A and B.

The electrons are being shared, with the positive nucleus of

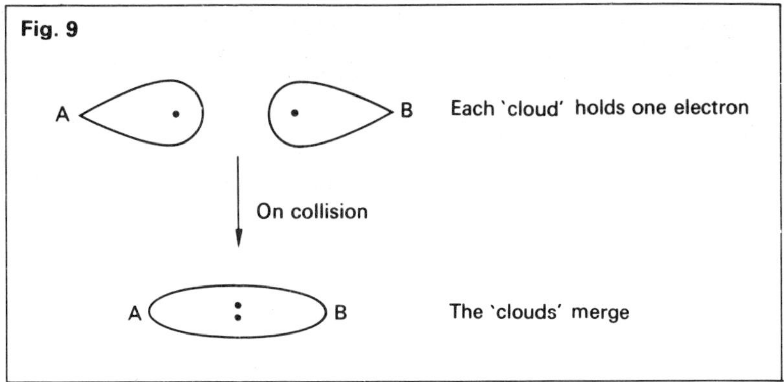

each atom attracting both electrons and therefore holding the atoms together.

The fate of the shared electrons

There are three possibilities, as shown in Fig. 10.

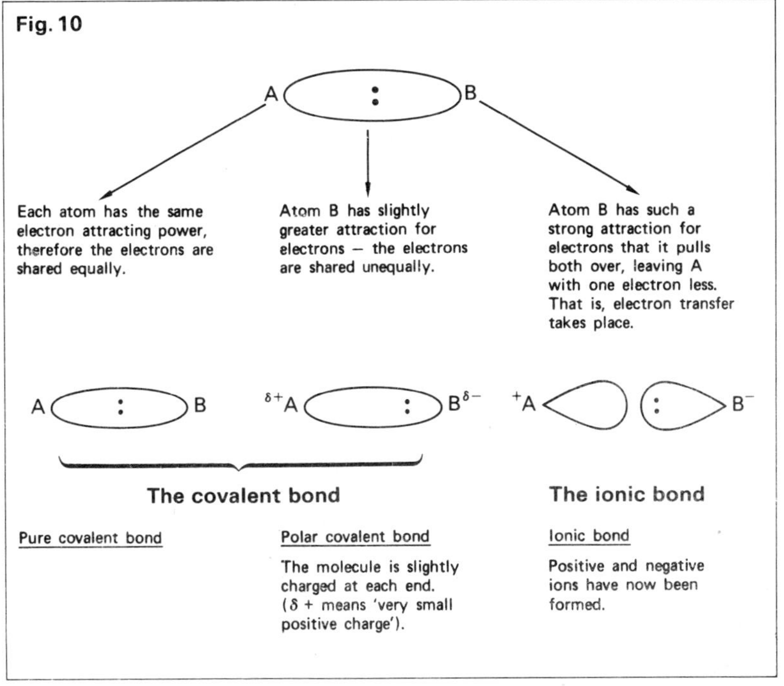

From this it can be seen that the type of bond formed depends on the electron attracting power of the atoms.

Electron attracting power

(a) Across each row: the nuclear charge (number of protons) is increasing, and the electron attracting power increases.

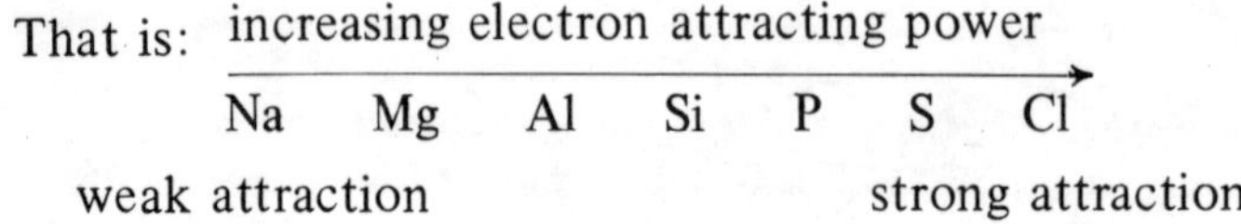

(b) Down each column: each row involves a new energy level, so the distance of the outer electrons from the nucleus increases greatly. Therefore, electron attracting power decreases.

That is: F Cl Br I

decreasing electron attracting power

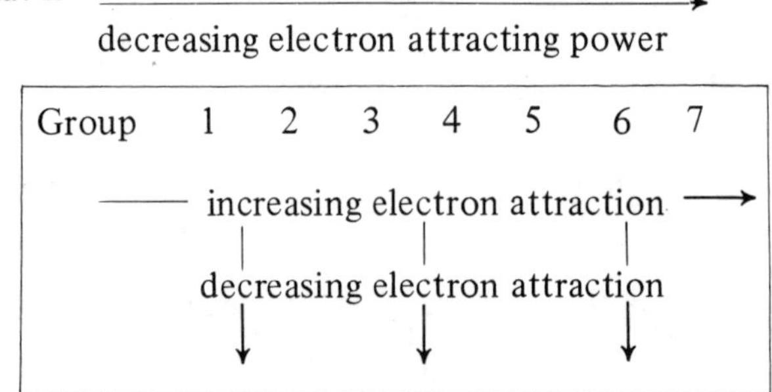

Thus fluorine (Group 7) has the greatest electron attracting power, and caesium (Group 1) has the lowest attracting power.

2.6 Will a bond be ionic or covalent?

If there is a big difference in electron attracting power of the two atoms, the bond will be ionic. If the difference is small, the bond will be covalent.

Ionic bond. Formed when metal atoms in Groups 1, 2 or 3 (weak electron attracting power) combine with non-metal atoms of Groups 5, 6 or 7 (strong electron attracting power).

Covalent bond (between atoms with no empty orbitals). Formed between non-metal atoms of Groups 4, 5, 6, 7, or hydrogen.

The covalent bond

Formation

Atoms have similar electron attracting power (Groups 4, 5, 6, 7, or hydrogen). Atoms collide, and their half-filled orbitals overlap.

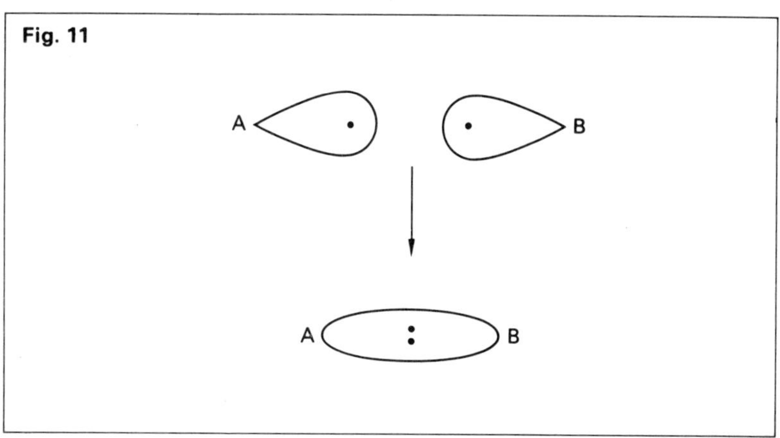

Fig. 11

The electrons remain shared, and both orbitals are filled by this process.

A *covalent bond* has been formed.

Examples

(a) *The chlorine molecule*

chlorine atom (one half-filled orbital)	+ chlorine atom (one half-filled orbital)		chlorine molecule
$|\overline{\underline{Cl}}\cdot$	$\cdot\overline{\underline{Cl}}|$	⟶	$|\overline{\underline{Cl}} - \overline{\underline{Cl}}|$
			'shared pair' of electrons

(b) *The oxygen molecule*

⟨O: The oxygen atom has two half-filled orbitals, and these can overlap with two half-filled orbitals from another oxygen atom, forming a *double bond*.

⟨O: :O⟩ ⟶ ⟨O═O⟩

(c) *The nitrogen molecule*

$$|\dot{\underset{\cdot}{N}}\cdot \quad \cdot\dot{\underset{\cdot}{N}}| \longrightarrow |N \equiv N|$$

A *triple bond* has been formed.

(d) *Tetrachloromethane* (carbon tetrachloride)

$$\cdot\dot{\underset{\cdot}{C}}\cdot \qquad \cdot\overline{\underline{Cl}}|$$

The atoms of carbon and chlorine are shown. For the carbon to achieve a full outer energy level (4 filled orbitals) 4 chlorine atoms are required for every one carbon atom. The result is:

$$\begin{array}{ccccc} & & |\overline{Cl}| & & \\ & & | & & \\ |\overline{\underline{Cl}} & — & C & — & \overline{\underline{Cl}}| \\ & & | & & \\ & & |\underline{Cl}| & & \end{array}$$

The carbon has 4 shared pairs, and each chlorine 1 shared pair of electrons.

Polar molecules

Since atoms of different elements have different electron attracting powers, there will be unequal sharing of electrons in most cases. In other words, the bond will be *polar* (it will have very small electric charges at each end).

Examples

(a) *The hydrogen chloride molecule*

hydrogen atom (one half-filled orbital)	+	chlorine atom (one half-filled orbital)
H·	↓	$\cdot\overline{\underline{Cl}}|$

hydrogen chloride

$$H—\overline{\underline{Cl}}|$$

'shared pair' of electrons.

Since chlorine has a stronger attraction for electrons than hydrogen, the bond will be polar:

$$^{\delta+}H - Cl^{\delta-}$$

(b) *Water molecule*

H· ·Ọ|

The oxygen atom has 2 half-filled orbitals and requires 2 shared pairs, whereas the hydrogen atom has only 1 half-filled orbital and requires one shared pair.
If the oxygen is to have its 2 shared pairs, 2 hydrogen atoms are required.

Because the oxygen has a stronger electron attraction, the water molecule will be polar:

$$H^{\delta+} \quad {}^{\delta-}O \quad H_{\delta+}$$

The shapes of molecules

Since combining atoms tend to arrange their 4 orbitals in a 3-dimensional tetrahedral structure, we should be able to predict the shape of many molecules. Some examples are given below.

(a) *Tetrachloromethane* (carbon tetrachloride)

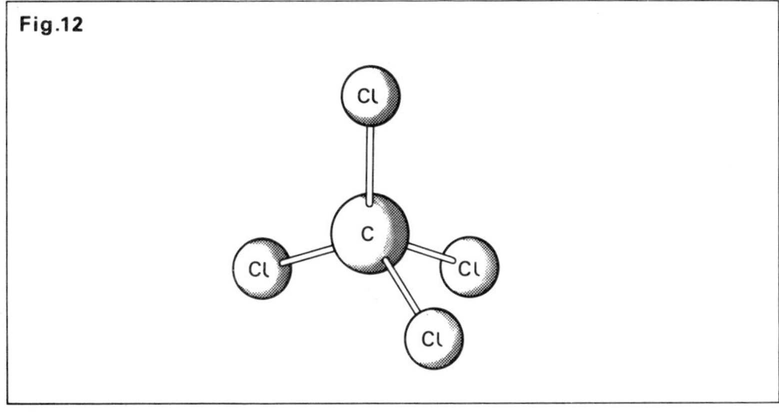

Fig.12

(b) *Water*

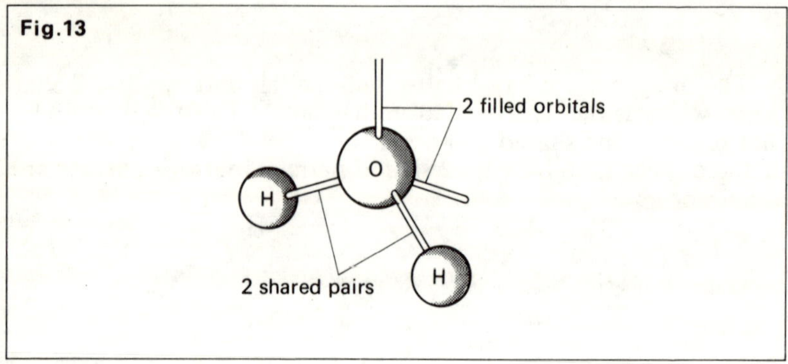

A water molecule should have the shape $\begin{matrix} & H \\ O & \\ & H \end{matrix}$

2.7 The ionic bond

Formation

Combining atoms have a big difference in electron attracting power. Atoms collide, and their half-filled orbitals overlap.

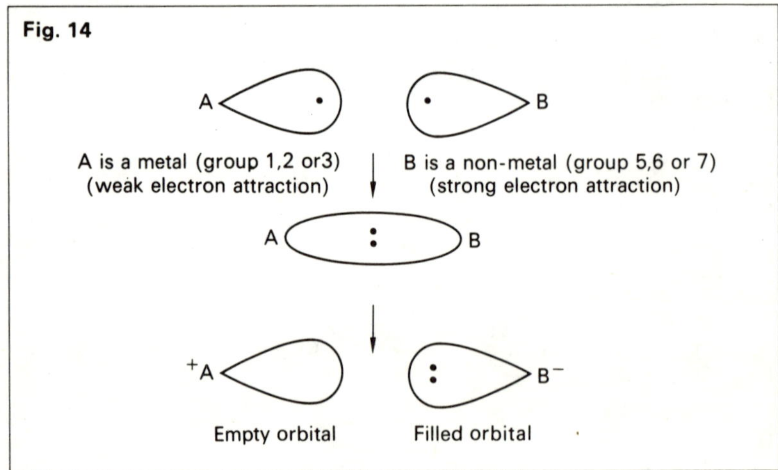

An electron has been pulled from A completely over to B, and *ions* have been formed. B has gained an electron, becoming negatively charged, while A has lost an electron, becoming positively charged.

Examples

(a) An example of a Group 1 metal with a Group 7 non-metal is sodium chloride.

Sodium, Na, 2.8.1.
1 half-filled orbital.

Chlorine, Cl, 2.8.7.
1 half-filled orbital.
3 filled orbitals.

Each chlorine atom can pull the outer electron from a sodium atom.

1 electron

Na, 2.8.1 → Cl, 2.8.7

The result is sodium chloride:

Na^+, 2.8.
Complete outer energy level – the electron in the orbital of the third energy level has disappeared.

Cl^-, 2.8.8.
Complete outer energy level – 4 filled orbitals.

(b) An example of a Group 2 metal with a Group 7 non-metal is calcium chloride.

Calcium, Ca, 2.8.8.2.
(2 half-filled orbitals)

Chlorine, Cl, 2.8.7.
(1 half-filled orbital)

Each chlorine atom can only pull 1 electron from the calcium, and, since the calcium will have a complete outer energy level if it loses 2 electrons, then 2 chlorine atoms are required for each calcium.

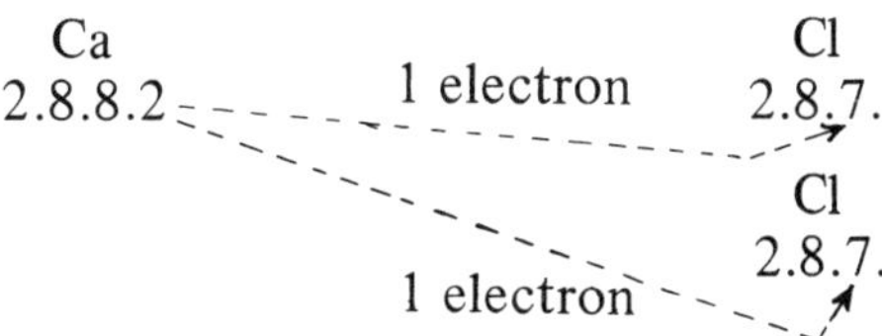

Two electrons are transferred, one to each chlorine atom. The result is calcium chloride:

Ca^{2+} 2.8.8. Cl^- 2.8.8.
Cl^- 2.8.8.

Shapes in ionic compounds

Since ions exist separately, there are no molecules in ionic compounds. For example, a crystal of sodium chloride is composed of a large number of positive and negative ions, held together by electrostatic forces.

Fig. 15 The sodium chloride crystal

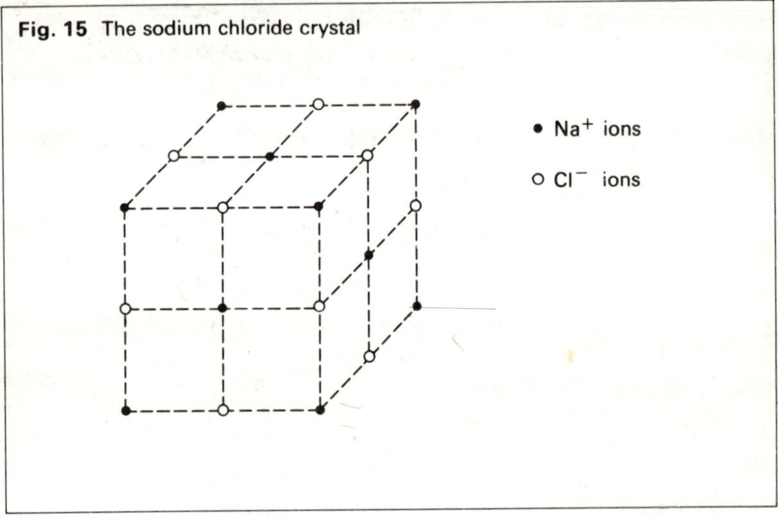

From this we see that the sodium chloride crystal takes up a cubic shape.

In other ionic compounds the ions are held together in a similar manner. However, they may take up different patterns. In other words, differently shaped crystals can be formed.

Some revision questions.

You should try to answer each question and then check your answer by referring to the section indicated in brackets after the question.

1 What is an electrolyte? (Section 2.1).
2 (a) Describe what happens to electrons in the formation of (i) the covalent bond, (ii) the ionic bond.
(b) Draw the three-dimensional shape of the following: (i) a water molecule, (ii) a molecule of ammonia (NH_3), (iii) part of the sodium chloride crystal.
(c) Why can we talk about molecules of water, but not about molecules of sodium chloride? (Section 2.4 to 2.7.)

1. An electrolyte is a substance which will conduct electricity when dissolved in water or when in a molten state. These substances are decomposed by electricity An electrolyte is usually a soln of an acid, alkali, salt or molten salt.

2 (a) (i) covalent bond : electrons equally shared between the atoms. IN PAIRS
(ii) Ionic bond : electrons completely transferred to one atom.

(b) (i)

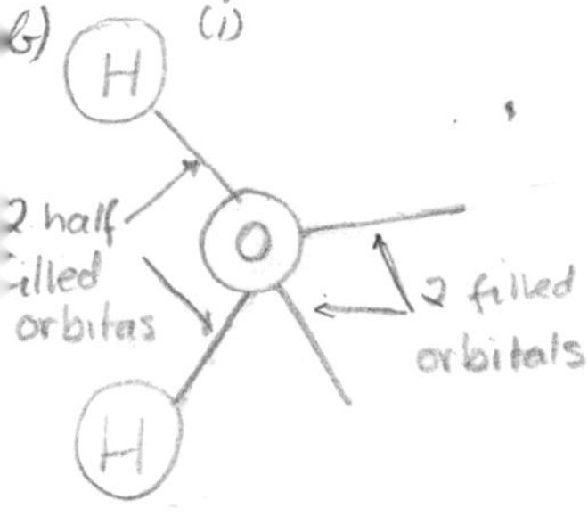

H
O
H
2 half filled orbitas
2 filled orbitals

(ii)

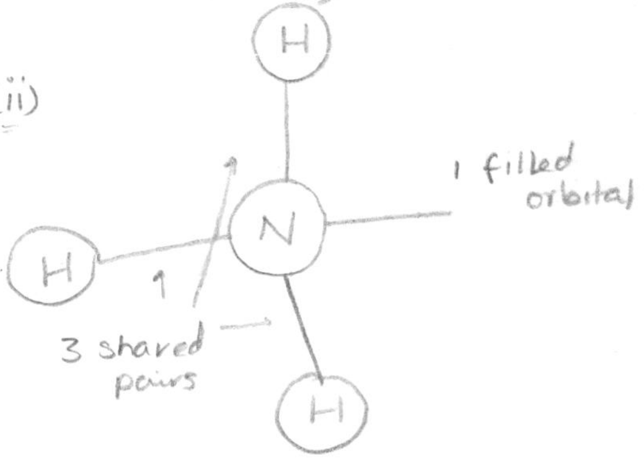

H
N
H
H
1 filled orbital
3 shared pairs

(iii)

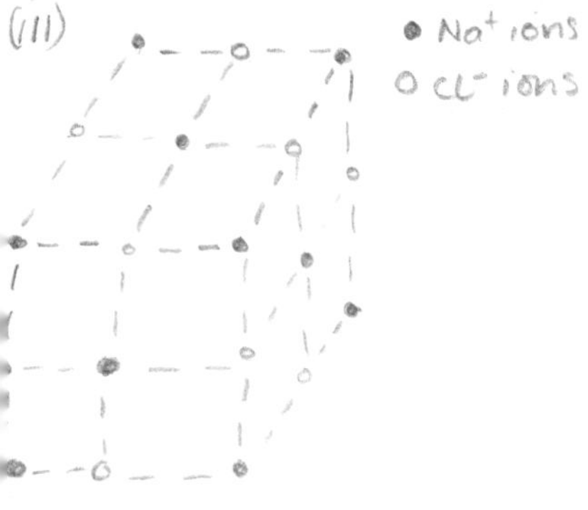

● Na^+ ions
○ Cl^- ions

(c) Molecules have no charge. Sodium chloride is charged and therefore can not be called a molecule.

3

Ions

3.1 Evidence for the existence of ions – electrolysis

Experiments in chapter 2 showed that salts conducted electricity when in solution in water, or in the molten state. When in the solid state, no conduction took place.

We suggested that since the metal (e.g. copper in copper chloride solution, and lead in molten lead bromide) was always deposited on the negative electrode (the cathode), then the metal particles in these compounds must be positively charged.

In the same way, we suggested that the non-metal particles in these salts had a negative charge.

On the evidence, it would therefore seem likely that ions exist, the salts conducting when in solution in water or in the molten state when the ions are free to move, but not conducting in the solid state when the ions are held in position.

3.2 Colour of compounds in solution

On looking at the colour of compounds in solution it can be seen that, for example, most potassium compounds are colourless, but potassium chromate is yellow. This suggests that the yellow is caused by the chromate ions. Since other chromate compounds are usually yellow, then it would seem reasonable that chromate ions are yellow.

On looking at other compounds in solution, it can be seen that:
dichromate ions are orange;
copper ions are blue;
nickel ions are green.
However, many ions are colourless (for example, sodium ions, potassium ions, chloride ions, sulphate ions, etc.).

3.3 Seeing ions move

If copper dichromate solution contains positively charged, blue copper ions and negatively charged, orange dichromate ions, then if an experiment is set up in a U-tube as in Fig. 16, we should be able to see the ions move. This is shown by a blue colour developing round the negative cathode, and an orange colour developing round the positive anode.

Fig. 16

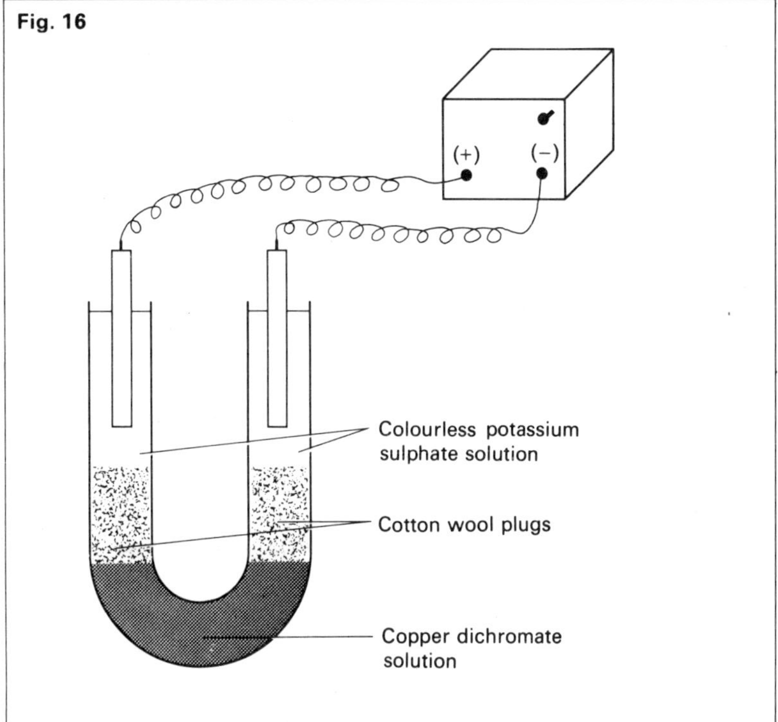

3.4 Large ions

Some ions are found which contain more than one element; the formula and names of these ions can be found in the data tables at the end of this book.

Example

The formula for the sulphate ion, SO_4^{2-}, means that 1 sulphur and 4 oxygen atoms are bonded covalently to form a group which is only stable when it has gained 2 extra electrons, giving an overall charge of '2–'.

A revision question

How can the electrolysis of copper chloride solution or molten lead bromide indicate that ions exist?

(Check your answer by referring to section 3.1.)

In a solution of copper chloride or lead bromide, the copper or lead was deposited on the negative electrode (cathode), therefore the metal must be positively charged. The non-metal particles have a negative charge

The result is

Pb^{2+}

Colours of compounds in sol^n

Chromate ions - yellow
Nickel ions - green
Dichromate ions - orange
Copper ions - blue

4

Formulae and the mole

4.1 Writing formulae

A formula is a neat, shorthand way of describing a substance. It should show not only the atoms present in a compound but the relative numbers of each atom present.

Let us look at previous ionic and covalent formulae obtained in 2.6 and 2.7.

Example – Water
Oxygen has 2 half-filled orbitals, and hydrogen has one half-filled orbital. Therefore 2 atoms of hydrogen are required for each atom of oxygen, and the formula is H_2O.

The following table compares a number of formulae.

	Compound	Elements	Group of Periodic Table	Number of half-filled orbitals before combining	Formula
covalent	hydrogen chloride	H Cl	1 7	1 1	HCl
covalent	water	H O	1 6	1 2	H_2O
covalent	tetrachloro-methane	C Cl	4 7	4 1	CCl_4
ionic	sodium chloride	Na Cl	1 7	1 1	NaCl
ionic	calcium chloride	Ca Cl	2 7	2 1	$CaCl_2$

We have written all the formulae above in the same way, the small numbers referring to the element just in front of the number. For example:

H_2O has 2 atoms of hydrogen to 1 atom of oxygen.
CCl_4 has 1 atom of carbon to 4 atoms of chlorine.

4.2 Valency number

For writing formula, the number of half-filled orbitals can be referred to as the combining number, or *valency number* of an atom.

Obtaining the valency number to write formulae

Let us compare the valency number (number of half-filled orbitals) of the atoms in Groups 1 to 7 of the Periodic Table, with the ionic charge given in the table *Properties of the Elements* at the end of this book.

Atoms of Group	**1 Li**	**2 Be**	**3 B**	**4 C**	**5 N**	**6 O**	**7 F**
Number of half-filled orbitals = valency number	1	2	3	4	3	2	1
Ionic charge	1+	2+	3+	4+	3–	2–	1–

The valency number and ionic charge are *numerically* the same.

If the ionic charge is used to obtain the valency number of atoms in Groups 1 to 7, then it can also be used for atoms not in these groups, and for large ions. (The ionic charge, and hence valency number, of the large ions being obtained from the data tables at the end of the book.)

eg.	**Atom or Group**	**Zn**	**Ag**	**SO_4**	**NO_3**
	Valency number	2	1	2	1

Using the valency number, we can now write the formula of many compounds.

Examples

	symbols	valency number (obtained from data tables)	formula
sodium chloride	Na Cl	1 1	NaCl
calcium chloride	Ca Cl	2 1	$CaCl_2$
magnesium nitride	Mg N	2 3	Mg_3N_2

	symbols	valency number (obtained from data tables)	formula
tetrachloromethane	C Cl	4 1	CCl_4
carbon dioxide	C O	4 2	CO_2
sodium sulphate	Na SO_4	1 2	Na_2SO_4
ammonium carbonate	NH_4 CO_3	1 2	$(NH_4)_2CO_3$
zinc oxide	Zn O	2 2	ZnO

4.3 Ions of variable charge

Some metal atoms can form ions of variable charge. In such cases, the charge (and therefore valency number) is indicated by Roman numerals after the metal.

eg. Copper (I) Oxide, Cu_2O ; Copper (II) oxide, CuO
Iron (II) oxide, FeO ; Iron (III) oxide, Fe_2O_3

Practice makes perfect

The only way to master formulae, is to try writing formulae for as many compounds as you can.

4.4 Structural formulae of compounds

Sometimes it is useful to show the structure of compounds and to do this we must decide whether the compound is ionic or covalent.

General rules

Ionic bond	positive ions	metals or NH_4^+ (ammonium ion)
	negative ions	non-metals of groups 5, 6, 7, or non-metal groups such as sulphate (SO_4^{2-}), nitrate (NO_3^-), etc.
Covalent bond	non-metals of groups 4, 5, 6, 7 or H.	

Structural formulae of ionic compounds – ionic formulae

To write an ionic formula, look up the ionic charges in the data tables and write the formula as on the next page.

Compound	formula	structural (or ionic) formula
sodium chloride	$NaCl$	$Na^{+}Cl^{-}$
calcium chloride	$CaCl_2$	$Ca^{2+}(Cl^{-})_2$
magnesium nitride	Mg_3N_2	$(Mg^{2+})_3(N^{3-})_2$
sodium sulphate	Na_2SO_4	$(Na^{+})_2SO_4^{2-}$
ammonium carbonate	$(NH_4)_2CO_3$	$(NH_4^{+})_2CO_3^{2-}$

For ions of variable charge, the Roman numerals tell us the ionic charge.

Examples: Copper (I) oxide Cu_2O, $(Cu^{+})_2O^{2-}$
Iron (III) oxide Fe_2O_3, $(Fe^{3+})_2(O^{2-})_3$

Structural formulae of covalent compounds

We know that:
Number of half-filled orbitals = charge on ion in tables
= Valency number.

Knowing the valency number of the atoms concerned tells us the number of shared pairs of electrons that each atom must have, and having written the formula, we can draw the structure to show the bonds between the atoms.

compound	symbols	valency number = no. of shared pairs	formula	structural formula
hydrogen chloride	H Cl	1 1	HCl	H − Cl
water	H O	1 2	H_2O	H–O–H (bent: O above, H on each side)
tetrachloromethane	C Cl	4 1	CCl_4	C bonded to four Cl (above, left, right, below)
carbon dioxide	C O	4 2	CO_2	O=C=O
ammonia	N H	3 1	NH_3	H—N—H, with a third H bonded below N

Remember that for covalent compounds formed with single bonds, we can draw the shape in 3-dimensions (see 2.6).

4.5 Chemical formulae deduced by experiment

We have so far deduced chemical formulae by theory, based on atomic structure. Originally, however, formulae were deduced experimentally by analysis or synthesis, and a knowledge of atomic weights (found from tables).

Experiments can be carried out in class to demonstrate this.

Examples

1 When 19.9 g of copper (II) oxide was reduced to copper, it was found that 15.9 g of copper was formed. Find the formula of the copper oxide.
We deduce that 15.9 g copper combined with 4 g oxygen,
therefore 1 g copper combined with $\frac{4}{15.9}$ g oxygen
The atomic weight of copper = 63.5,
therefore 63.5 g copper combined with $\frac{4}{15.9} \times 63.5$ g oxygen
$= 16$ g oxygen
But the atomic weights are Cu = 63.5; 0 = 16.
Weights present in the compound are Cu = 63 g; 0 = 16 g.
Therefore the simplest formula (empirical formula) of this compound is CuO.

2 20.7 g lead were dissolved in nitric acid, and excess sodium iodide solution added. The lead iodide precipitate obtained was filtered, dried and weighed. Its weight was found to be 46.1 g. Find the formula of the lead iodide.
20.7 g lead reacted with (46.1 − 20.7) = 25.4 g iodine.
Therefore 1 g lead reacted with $\frac{25.4}{20.7}$ g iodine.
the atomic weight of lead = 207,
therefore 207 g lead reacted with $\frac{25.4}{20.7} \times 207$ g iodine

$$= \frac{254}{207} \times 207$$

$$= 254 \text{ g iodine}$$

But the atomic weights are: lead 207; iodine 127.
Weights present in the compound are: lead 207 g; iodine 254 g.
Therefore the empirical formula is <u>PbI_2</u>.

4.6 Formula mass and percentage composition

The formula mass (formula weight) is obtained by adding together the masses (weights) of the atoms in the formula.

Copy out and complete the following table.

Compound	Formula	Formula mass (formula weight)	Gram formula mass (weight)
water	H_2O	(2 x 1) + 16 = 18 amu	18 g
carbon dioxide	CO_2	12 + (2 x 16) = 44 amu	44 g
ammonium sulphate	$(NH_4)_2SO_4$	(14 + 4) 2 + 32 + (4 x 16) = 132 amu	
sodium carbonate	Na_2CO_3		
sodium chloride			
sodium nitrate			
carbon tetrachloride			
calcium carbonate			

Percentage composition (by mass)

This is best explained by giving some examples.

(a) To find the percentage of hydrogen in water, H_2O.
The formula mass (formula weight) = 18 amu.
That is, in every 18 amu of water 2 amu are hydrogen and 16 amu are oxygen.
Provided we keep all units the same, we can use grams, kg, etc. in place of amu.
Therefore, in every 18 g of water there are 2 g hydrogen and 16 g oxygen.
Therefore % of hydrogen in water is $\frac{2}{18} \times 100 = 11.1\%$.

(b) To find the percentage of carbon in carbon dioxide, CO_2.
The formula mass = 44 (as above).
Therefore, in every 44 g of CO_2 there are 12 g of carbon.
Therefore % carbon in carbon dioxide is $\frac{12}{44} \times 100 = 27.3\%$.

4.7 The Mole

A mole of a substance is the formula mass expressed in grams. That is, 1 mole = 1 gram formula mass.

1 mole of any substance contains the same number of 'formulae'. The number of 'formulae' in a mole is very large and is in fact about 600 000 000 000 000 000 000 000 (six hundred thousand million million million).

Fig. 17

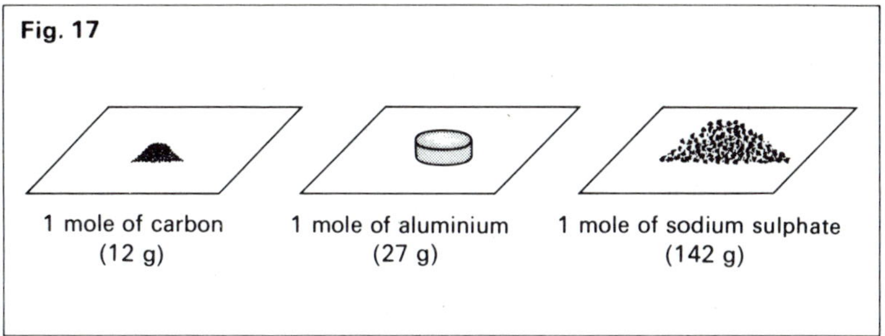

Why do we use the mole?

In theory we could make a molecule of water by reacting 2 atoms of hydrogen and 1 atom of oxygen.

In practice this is impossible, and we must react very large numbers of atoms or molecules together.

We therefore take a formula to mean *one mole* of a material, and can say that, in the case of water (H_2O), 1 *mole* of water (18 g) is formed from 2 *moles* of hydrogen atoms (2 g) and one *mole* of oxygen atoms (16 g).

Also, in the case of sodium sulphate (Na_2SO_4), one *mole* of sodium sulphate contains 2 *moles* of sodium ions and one *mole* of sulphate ions.

Molarity

Since reactions are very often carried out in solution in water, we must be able to discuss concentrations of solutions.

A molar solution is a solution containing 1 mole of a substance in 1 litre of solution.

e.g. NaCl. 1 mole weighs (23 + 35.5) g = 58.5 g.
Therefore, M NaCl contains 58.5 g in each litre of solution.
2M NaCl contains 117 g in each litre of solution.

Some Revision questions

You should try to answer each question and then check your answer by referring to the section indicated in brackets after the question.

1 Use the Periodic Tables at the end of this book to help you with this question.

(a) Write the formula for each of the following compounds: (i) potassium chloride; (ii) calcium sulphate; (iii) phosphorus chloride; (iv) copper (II) nitrate; (v) hydrogen sulphide. (Sections 4.1 to 4.3.)

(b) Take each of the above in turn, and state whether the compound is ionic or covalent. (Section 4.4.)

(c) Rewrite the formula, this time indicating the structure. In other words, write the ionic formula, or the covalent structural formula. (Section 4.4.)

2 Try the following problems.

(a) 5.6 g iron was found to react with 1.6 g oxygen. Find the formula of the iron oxide.

(b) 1.2 g magnesium was burned, and formed 2 g of magnesium oxide. What is the formula of the magnesium oxide?

(c) 1.6 g of sulphur was burned in air. The weight of sulphur dioxide formed was 3.2 g.
What is the formula of the sulphur dioxide?
(Section 4.5.)

3 Calculate the percentage of nitrogen in each of the following compounds and hence, since nitrogen compounds are important as fertilisers, pick out the best fertiliser.

(i) Ammonium sulphate, (ii) ammonium nitrate, (iii) ammonia (NH_3).
(Section 4.6.)

4 What is (i) a mole, (ii) a molar solution? (Section 4.7.)

5 How many moles are

(a) 112 g of iron;
(b) 150 g of calcium carbonate;
(c) 14.2 g of sodium sulphate;
(d) 16 g of oxygen gas (O_2)? (Section 4.7.)

6 How many grams of each of the following substances would you weigh out in order to make up the following solutions:

(a) 1 litre of 2M sodium chloride;

(b) 1 litre of $\frac{M}{2}$ sodium sulphate;
(c) 100 cm³ (0.1 litre) of M silver nitrate;
(d) 100 cm³ of $\frac{M}{10}$ sodium hydroxide? (Section 6.7.)

7 How many moles are contained in:
(a) 1 litre of 3M potassium hydroxide;
(b) 100 cm³ of M ammonium chloride;
(c) 50 cm³ of 0.5M sulphuric acid;
(d) 100 cm³ of 0.1M sodium chloride? (Section 4.7.)

8 What is the molarity of the following solutions:
(a) 6 moles in 2 litres of solution;
(b) 1 mole in 500 cm³ of solution;
(c) 0.1 moles in 100 cm³ of solution;
(d) 2 moles in 4 litres of solution? (Section 4.7.)

5

Equations

An equation is a means of summing up a chemical reaction.

5.1 Unbalanced equations

These can be used as a shorthand method of stating the change taking place in a reaction.
For example: $MgO + HCl \longrightarrow MgCl_2 + H_2O$

In other words, magnesium oxide plus hydrochloric acid forms magnesium chloride plus water.

This equation tells us *nothing* about quantities used or the conditions of the experiment. However, *this unbalanced equation is often all that is required.*

5.2 Balanced equations

These give us information about quantities used.

For example:	MgO +	$2HCl$	$\longrightarrow$	$MgCl_2$ +	H_2O
	1 mole	2 moles		1 mole	1 mole

This tells us that for a complete reaction, we need 2 moles of hydrochloric acid for every 1 mole of oxide. (See Fig. 18.)

5.3 Using equations in chemical calculations

If we write a balanced equation, we can use this equation to predict the quantities of a material required for an experiment, or to predict what quantities of the products would be obtained.

Examples

(a) What weight of carbon dioxide would you expect from adding

Fig. 18

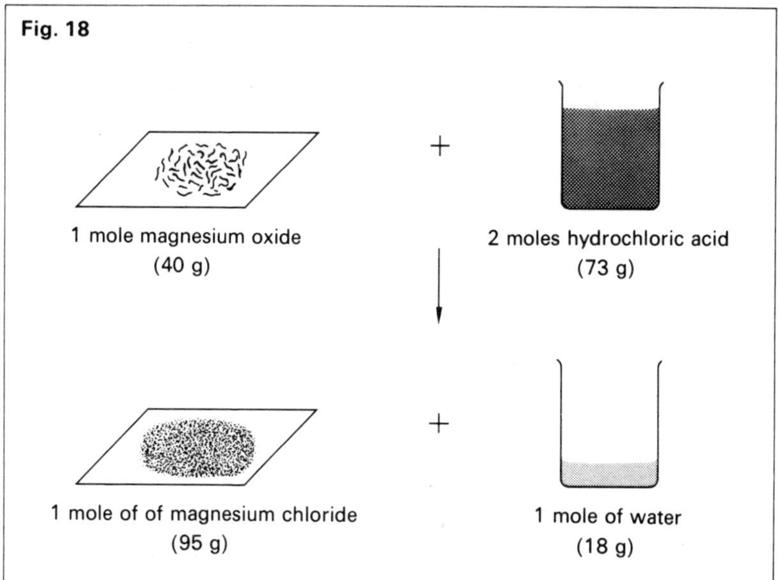

excess hydrochloric acid to 5 g of calcium carbonate?

Balanced equation:

$$\underset{1\text{ mole}}{CaCO_3} + 2HCl \longrightarrow CaCl_2 + H_2O + \underset{1\text{ mole}}{CO_2}$$

From the equation, we see that 1 mole of $CaCO_3$ gives 1 mole of CO_2.

That is, (40+12+48) g $CaCO_3$ gives (12+32) g CO_2

or 100 g $CaCO_3$ gives 44 g CO_2

Therefore, 5 g $CaCO_3$ gives $\frac{44 \times 5}{100}$ g CO_2

$= 2.2$ g CO_2

(b) How many moles of hydrochloric acid would be required to react with 10 g magnesium oxide?

Balanced equation:

$$\underset{1\text{ mole}}{MgO} + \underset{2\text{ moles}}{2HCl} \longrightarrow MgCl_2 + H_2O$$

From the equation, we see that:

1 mole of MgO requires 2 moles of HCl
That is, (24+16) g of MgO requires 2 moles of HCl
or, 40 g of MgO requires 2 moles of HCl
Therefore, 10 g MgO requires $\frac{1}{2}$ mole of HCl
= 0.5 moles of HCl

Your teacher may give you some problems to try for yourself.

5.4 State equations

State equations give us more information about a reaction. The letters (s), (l), (g), or (aq) are used to tell us if the materials are solids, liquids, gases, or in solution in water (aqueous).

e.g. $MgO(s) + 2HCl(aq) \longrightarrow MgCl_2(aq) + H_2O(l)$

5.5 Ionic equations (with more information)

We can rewrite the balanced state equation above showing the ions which are present.

$$Mg^{2+}O^{2-}(s) + 2H^+Cl^-(aq) \longrightarrow Mg^{2+}(Cl^-)_2(aq) + H_2O(l)$$

Since ionic compounds in solution are present as free ions, we could expand this equation even further, to give:

$$Mg^{2+}O^{2-}(s) + 2H^+(aq) + 2Cl^-(aq) \longrightarrow Mg^{2+}(aq) + 2Cl^-(aq) + H_2O(l)$$

Here we see that some species have remained unchanged throughout the reaction – in this case the Cl^- ions have remained in solution throughout.

These ions have not taken any part in the reaction, and are therefore called *spectator ions.*
The spectator ions can be omitted from the equation:

$$Mg^{2+}O^{2-}(s) + 2H^+(aq) \longrightarrow Mg^{2+}(aq) + H_2O(l)$$

This equation tells us that 1 mole of magnesium oxide added to

2 moles of hydrogen ions (in *any* dilute acid), forms 1 mole of magnesium ions in solution and one mole of water.

Another example is that of dilute sulphuric acid and sodium hydroxide solution reacting to form sodium sulphate solution and water.

Unbalanced equation: $H_2SO_4 + NaOH \longrightarrow Na_2SO_4 + H_2O$

Balanced state equation:
$H_2SO_4(aq) + 2NaOH\ (aq) \longrightarrow Na_2SO_4(aq) + 2H_2O(l)$

Ionic equation:
$(H^+)_2SO_4^{2-}(aq) + 2Na^+OH^-(aq) \longrightarrow (Na^+)_2SO_4^{2-}(aq) + 2H_2O(l)$

In order to see more clearly which ions are the spectator ions, we can expand the equation to show free ions in solution.

i.e. $2H^+(aq) + SO_4^{2-}(aq) + 2Na^+(aq) + 2OH^-(aq)$
$\longrightarrow 2Na^+(aq) + SO_4^{2-}(aq) + 2H_2O(l)$

We see that the $Na^+(aq)$ and the $SO_4^{2-}(aq)$ ions are spectator ions, and if these are omitted the equation could now read:

$$2H^+(aq) + 2OH^-(aq) \longrightarrow 2H_2O(l)$$
$$\text{or}\quad H^+(aq) + OH^-(aq) \longrightarrow H_2O(l)$$

That is, the reaction taking place is only between the hydrogen ions of the acid and the hydroxide ions of the alkali, forming water.

5.6 The formula of acids

The formula of hydrogen chloride has earlier been written as HCl, and it has been thought of as a covalent material. However, in 2.1 you learned that acids such as hydrochloric acid were electrolytes and therefore contained free ions.

When hydrogen chloride, HCl (g), dissolves in water hydrochloric acid, HCl(aq), is obtained and ions are produced, $H^+Cl^-(aq)$.

This is explained in chapter 7, but meantime we can consider all acids in solution in water as containing free hydrogen ions, $H^+(aq)$.

Dilute hydrochloric acid, $HCl(aq)$, consists of $H^+Cl^-(aq)$.
Dilute sulphuric acid, $H_2SO_4(aq)$, consists of $(H^+)_2SO_4{}^{2-}(aq)$.
Dilute nitric acid, $HNO_3(aq)$, consists of $H^+NO_3{}^-(aq)$.

Some revision questions

You should try to answer each question and then check your answer by referring to the section indicated in brackets after the question.

1 When solid calcium carbonate is added to dilute hydrochloric acid, carbon dioxide is given off, and the calcium chloride and water are formed.

(a) Write an unbalanced equation for the above reaction.
(b) Write the balanced equation.
(c) How many moles of calcium carbonate would you have to use to get 0.2 moles of carbon dioxide? (Assume that there is sufficient acid for the reaction).
(d) What weight of calcium carbonate would this be?
(e) What weight of calcium chloride would be obtained?

(Section 5.1 to 5.3. For the formulae refer to 4.1 and 4.2.)

2 When zinc is added to copper (II) sulphate solution, copper metal is formed and the zinc goes into solution as zinc sulphate.

(a) Write the balanced state equation for this reaction.
(b) Write the ionic equation.
(c) Rewrite the ionic equation, this time omitting spectator ions. (Section 5.4 and 5.5.)

6

Activity and the electrochemical series

6.1 Revision of experiments with metals

(a) Burning metals in oxygen

Using the apparatus in Fig. 19, we can find an *order of activity* for the metals.

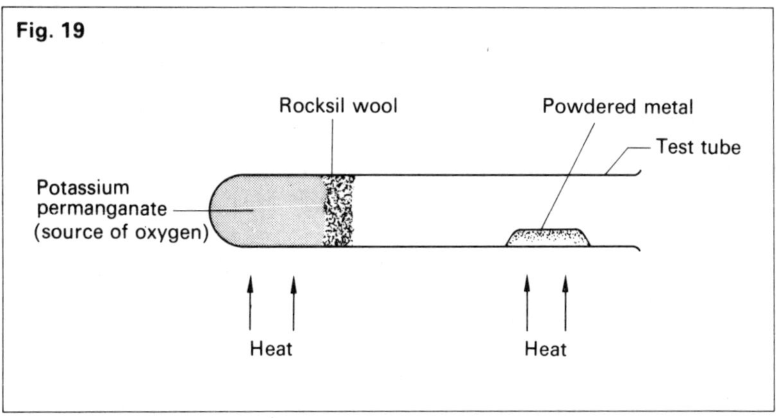

Fig. 19

Metal		Equation	Reactivity
Magnesium	:	$2Mg + O_2 \rightarrow 2MgO$	↑ most reactive
Aluminium	:	$4Al + 3O_2 \rightarrow 2Al_2O_3$	
Zinc	:	$2Zn + O_2 \rightarrow 2ZnO$	
Iron	:	$4Fe + 3O_2 \rightarrow 2Fe_2O_3$	
Lead	:	$2Pb + O_2 \rightarrow 2PbO$	
Copper	:	$2Cu + O_2 \rightarrow 2CuO$	least reactive

(b) Metals and water

By placing small pieces of metal in water (see Fig. 20), the following order of activity can be found (most reactive first);

Potassium	:	$2K + 2H_2O \rightarrow 2KOH + H_2$
Sodium	:	$2Na + 2H_2O \rightarrow 2NaOH + H_2$
Lithium	:	$2Li + 2H_2O \rightarrow 2LiOH + H_2$
Calcium	:	$Ca + 2H_2O \rightarrow Ca(OH)_2 + H_2$

Hydrogen and an alkali are formed.
Other metals react slowly, if at all, with water.

(c) Metals and acid

(Note: potassium, sodium, lithium and calcium are *never* added to acid.) If other metals are reacted with acids, the following order of activity is found (most reactive first):

Magnesium	:	$Mg(s) + 2H^+(aq) \rightarrow H_2(g) + Mg^{2+}(aq)$.	With dilute hydrochloric acid.
Aluminium	:	$2Al(s) + 6H^+(aq) \rightarrow 3H_2(g) + 2Al^{3+}(aq)$.	
Zinc	:	$Zn(s) + 2H^+(aq) \rightarrow H_2(g) + Zn^{2+}(aq)$.	
Iron	:	$Fe(s) + 2H^+(aq) \rightarrow H_2(g) + Fe^{2+}(aq)$.	
Tin	:	$Sn(s) + 2H^+(aq) \rightarrow H_2(g) + Sn^{2+}(aq)$.	
Lead	:	$Pb(s) + 2H^+(aq) \rightarrow H_2(g) + Pb^{2+}(aq)$.	with concentrated hydrochloric acid.
Copper	:	$Cu(s) + H^+(aq) \rightarrow$ no reaction	

By combining the results from these experiments, an *activity series* is built up.

Potassium	K	↑ most active
Sodium	Na	
Lithium	Li	
Calcium	Ca	
Magnesium	Mg	
Aluminium	Al	
Zinc	Zn	
Iron	Fe	
Tin	Su	
Lead	Pb	
Copper	Cu	least active

6.2 Displacement of one metal by another

Active metals can *displace* less active metals from their salts. For

Fig. 20

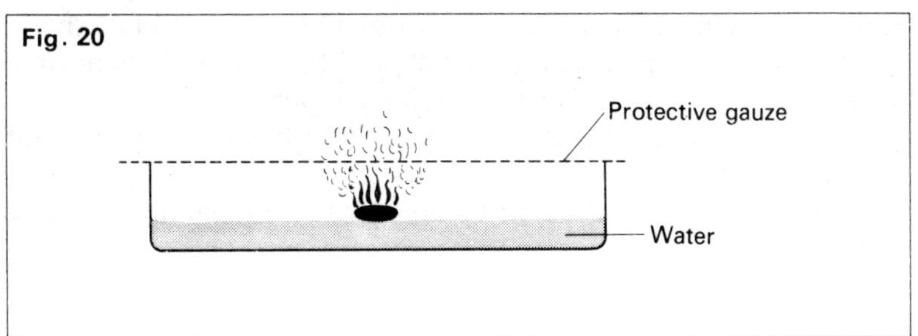

example, magnessium will react with copper sulphate solution to produce magnesium sulphate and copper:

$$Mg(s) + Cu^{2+}(aq) \rightarrow Mg^{2+}(aq) + Cu(s)$$

Similarly: $Fe(s) + Cu^{2+}(aq) \rightarrow Fe^{2+}(aq) + Cu(s)$

But: $Pb(s) + Mg^{2+}(aq) \not\rightarrow$ no reaction, since lead is less active than magnesium.

Similarly: $Mg(s) + Mg^{2+}(aq) \not\rightarrow$, $Cu(s) + Cu^{2+}(aq) \not\rightarrow$ } no reaction

Displacement, and reacting masses

From the *balanced* equation

$$Mg(s) + Cu^{2+}(aq) \rightarrow Mg^{2+}(aq) + Cu(s)$$

We see that 1 mole of magnesium should displace 1 mole of copper from Cu^{2+} solution.
That is, 24 g Mg should displace 63.5 g Cu from Cu^{2+} ions in solution.
That is, 1 g Mg should displace $\frac{63.5}{24} = 2.67$ g Cu.
This result may be verified by experiment.

6.3 Ion-electron equations, and electron transfer

The equation, $Mg(s) + Cu^{2+}(aq) \rightarrow Mg^{2+}(aq) + Cu(s)$ can be split up into 2 equations, describing the changes taking place. (These ion-electron equations can be obtained from the data tables at the end of this book.)

$Mg(s) \rightarrow Mg^{2+}(aq) + 2e$
$Cu^{2+}(aq) + 2e \rightarrow Cu(s)$

Since the magnesium atoms have formed magnesium ions, they must have lost 2 electrons per atom, and the copper ions must have picked up 2 electrons in forming copper atoms.

Electrons transfer must have taken place from the magnesium atoms to the copper ions.

This may be demonstrated by an experiment similar to that in Fig. 21.

Fig. 21

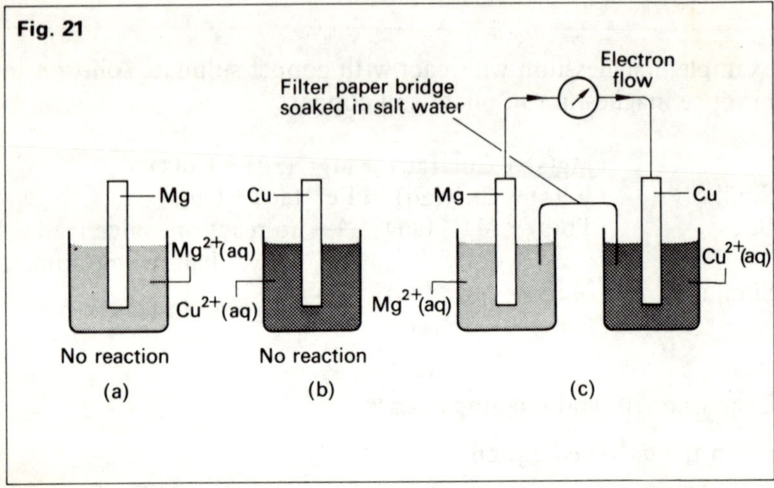

In Fig. 21(c), magnesium forms magnesium ions in solution, and fresh copper is deposited on the copper electrode.

6.4 Reduction of oxides

Increasing ease of reduction ↓

Oxide of	
Potassium Sodium Calcium Magnesium Aluminium Zinc	Not readily reduced to the metal.
Iron Tin Lead Copper	Reduced to the metal by the action of hydrogen or carbon on hot oxide.
Mercury Silver	Decomposed by heat alone.

Since the most active metal is the one which forms ions most readily (e.g. $K \rightarrow K^+ + e$), then the ions formed will be least likely to pick up electrons to reform the metal. (That is, $K^+ + e \rightarrow K$ will be difficult to achieve.) Thus, the metal oxides of metals low in the activity series are easier to reduce to the metals than the oxides of active metals.

6.5 Ease of forming ions in solution

All metals tend to form ions in solution. When zinc metal is placed in water, it tends to form zinc ions:
$Zn(s) \rightarrow Zn^{2+}(aq) + 2e$

The electrons are left on the zinc plate, giving it a negative charge (Fig. 22(a)).

Similarly, with copper, some copper ions will form and the metal will develop a negative charge due to electrons being left behind (Fig. 22(b)).

However, all metals do not form ions equally readily. In the above case, the zinc forms ions more readily than the copper. As a result, there will be a bigger 'build up' of electrons (that is, a higher electrical pressure or *potential*) on the zinc plate.

Fig. 22 (a)

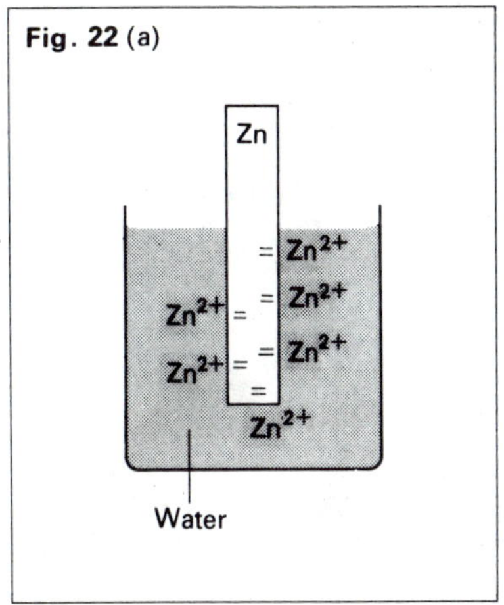

Fig. 22 (b)

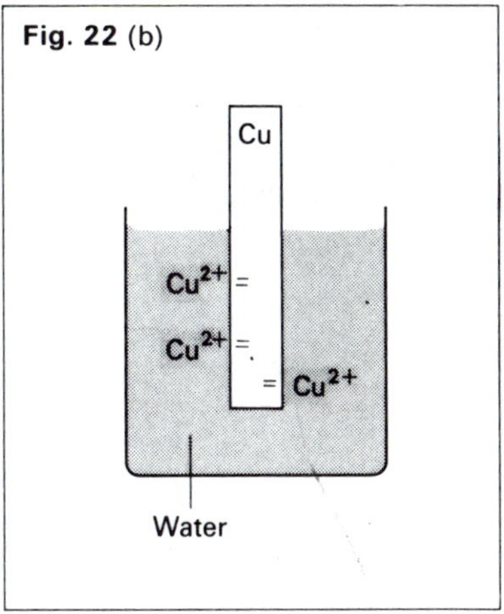

If the two strips are connected by a wire, the potential difference (voltage) between the plates forces electrons to flow from the zinc to the copper. A voltmeter can measure this potential difference (Fig. 23).

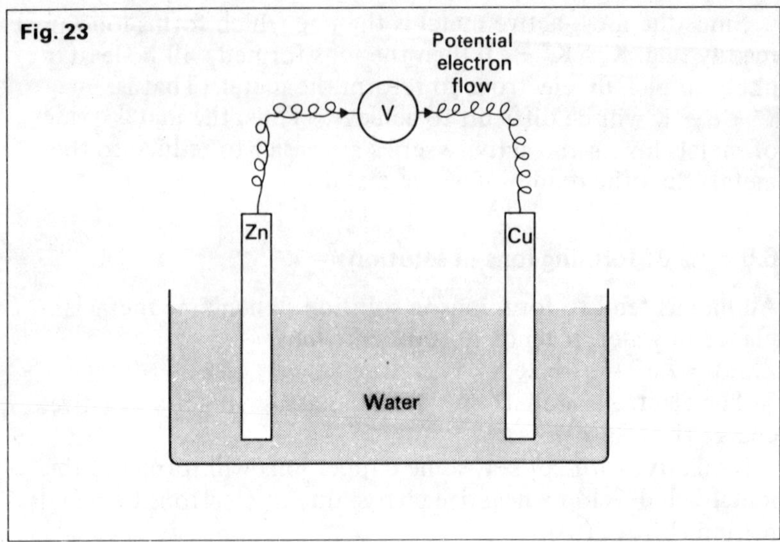

Fig. 23

The potential difference between different metal pairs can be measured. For our purposes, copper metal can be a convenient standard to use (although it is not used by chemists as the standard) and if the metals are arranged in order of voltage, the series formed is called the Electrochemical Series.

The Electrochemical Series.

↑		
Metals *most readily* forming $M^+(aq)$ ions	Lithium: $Li(s) \rightleftharpoons Li^+(aq) + e$	Metal ions *least readily* form the metal
	Potassium: $K(s) \rightleftharpoons K^+(aq) + e$	
	Calcium: $Ca(s) \rightleftharpoons Ca^{2+}(aq) + 2e$	
	Sodium: Na	
	Magnesium; Mg	
	Aluminium; Al	
	Zinc, Zn	
	Iron, Fe	
	Tin, Sn	Metal ions *most readily* form the metal
	Lead, Pb	
Metals *least readily* forming $M^+(aq)$	Copper: $Cu \rightleftharpoons Cu^{2+}(aq) + 2e$	
	Mercury: $Hg \rightleftharpoons Hg^{2+}(aq) + 2e$	

Note: the series is very similar to the activity series, *but* lithium is at the top in this series, and the positions of sodium and calcium are reversed.

The Electrochemical Series compares the *energy* involved in converting metal atoms into metal ions in solution.

The Reactivity Series compares the *rates* of reactions involving metals. In other words, lithium may take longer to react than potassium, but the total energy released on forming ions in solution is greater.

6.6 Oxidation and reduction – redox reactions

Oxidation used to be thought of as gain of oxygen, for example:

$2Mg + O_2 \rightarrow 2MgO$

The magnesium atoms have formed magnesium ions, and lost electrons in the process:

$Mg \rightarrow Mg^{2+} + 2e$

When magnesium burns in chlorine, magnesium chloride is formed:

$Mg + Cl_2 \rightarrow MgCl_2$

Here again the magnesium atoms have lost electrons and formed ions:

$Mg \rightarrow Mg^{2+} + 2e$

Since the same change has taken place in both these reactions (the magnesium has lost electrons and formed ions), the meaning of oxidation has been extended to cover all such changes. **Oxidation is the process of electron loss.**
(Remember: Loss of Electrons is Oxidation – **LEO.**)
The reverse process, electron gain, is called reduction. **Reduction is the process of electron gain.**
(Remember: Reduction is Electron Gain – **REG.**)

Since both electron loss (oxidation) and electron gain (reduction) must take place in the same reaction, an oxidation-reduction reaction is usually called a *redox* reaction.

We have already discussed the electron transfer from magnesium atoms to copper ions; the reactions taking place being:

$Mg(s) \rightarrow Mg^{2+}(aq) + 2e$ Oxidation
$Cu^{2+}(aq) + 2e^{-} \rightarrow Cu(s)$ Reduction

Cells can be set up to demonstrate electron transfer in other redox reactions. For example (Fig. 24):

$Fe^{2+} \rightarrow Fe^{3+} + e$
(iron (II) ions) (iron (III) ions) Oxidation

$I_2 + 2e \rightarrow 2I^{-}$
iodide ions Reduction

Fig. 24

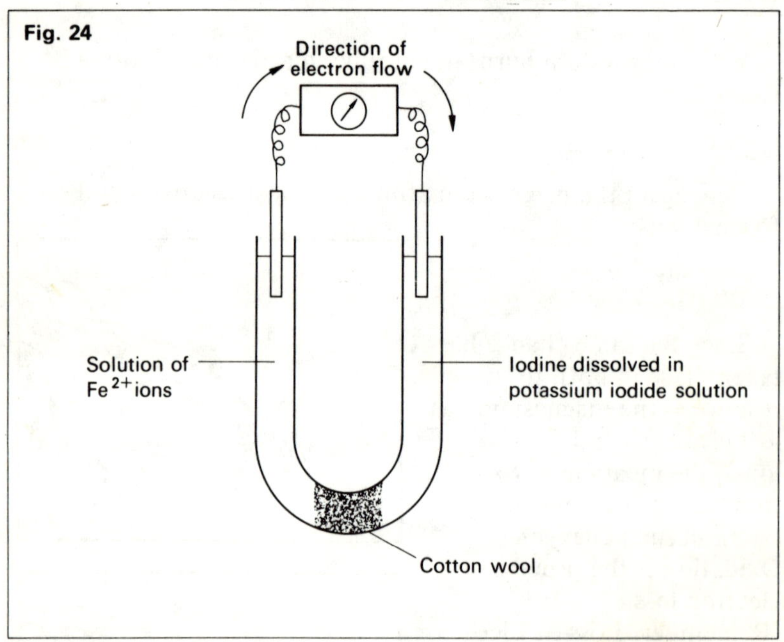

Note: Most of the ion-electron equations which you will use, can be obtained from the data tables Standard Reduction Electrode Potentials – at the end of this book.

Redox reactions in electrolysis

Let us look at the electrode reactions during electrolysis of a number of compounds.

	Cathode	Anode
1 Copper (II) chloride solution, $CuCl_2$	$Cu^{2+} + 2e \rightarrow Cu$	$2Cl^- \rightarrow Cl_2 + 2e$
2 Molten lead bromide, $PbBr_2$	$Pb^{2+} + 2e \rightarrow Pb$	$2Br^- \rightarrow Br_2 + 2e$
3 Extraction of aluminium from molten bauxite, Al_2O_3	$Al^{3+} + 3e \rightarrow Al$	$2O^{2-} \rightarrow O_2 + 4e$
	Electron gain, that is **reduction** at the cathode.	Electron loss, that is **oxidation** at the anode.

6.7 Corrosion

When metals react with their natural surroundings (air or water) they form ions, and in so doing lose electrons.

This process is often called corrosion.

Corrosion of iron

Experiments show that before iron will rust, both water and air must be present.

Since iron is the major metal used in the world, we must find efficient methods of protecting it.

Two common methods used are: (a) galvanising (zinc plating);
(b) tin plating.

Efficiency of galvanising and tin plating

Both galvanising and tin plating are effective in stopping corrosion of iron so long as the layers are complete. However, when the surfaces are scratched there is a considerable difference:

Galvanised iron – there is still little or no corrosion of the iron.
Tin plated iron – the iron now corrodes very quickly.

This suggests that the process of rusting (iron forming ions) may either be assisted or prevented, depending on what is connected to the iron.

The electrochemical nature of corrosion

When iron is connected to another metal, electrons will either flow to the iron, or from it, depending on the position of the other metal in the electrochemical series. Ferroxyl indicator (a mixture of phenolphthalein and potassium hexacyanoferrate III) is useful in investigations since in the presence of iron (II) ions it turns blue, and in the presence of hydroxide ions it turns pink.

Experiments similar to those in Fig. 25 can be carried out to find out more about the corrosion process.

Fig 25 Investigation of corrosion. Each U tube contains NaCl (aq) and ferroxyl indicator

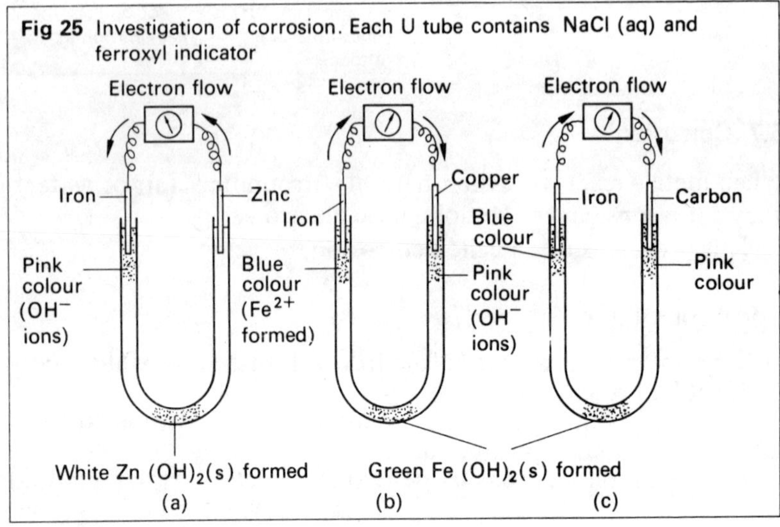

The following refers to Fig. 25.

(a) Iron and zinc

The zinc forms ions more readily than the iron, and so goes into solution as zinc ions.

$$Zn(s) \longrightarrow Zn^{2+}(aq) + 2e$$

The electrons are pushed on to the iron, which does not corrode. Instead, hydroxide ions form at the iron.

(b) Iron and copper

Iron forms ions more readily than copper.

$$Fe(s) \longrightarrow Fe^{2+}(aq) + 2e$$

Electrons flow from the iron to the copper, and the hydroxide ions form at the copper.

(c) Iron and carbon

Since carbon is used in the reduction of iron ore in the blast furnace, there is always carbon as impurity in iron, and it is important to study its effect. Once again the iron ionises.

$$Fe(s) \longrightarrow Fe^{2+}(aq) + 2e$$

The electrons flow from the iron to the carbon, and hydroxide ions form at the carbon.

We can now conclude from the experiment that if electrons are pushed on to the iron it will not form ions, and if the electrons are encouraged to leave the iron, it will form ions (corrode) quickly.

The normal rusting of iron

Due to carbon impurity, corrosion of iron is speeded up by the formation of an iron/carbon cell as in Fig. 25(c).

$$Fe \longrightarrow Fe^{2+} + 2e$$

The electrons flow to the carbon and hydroxide ions form at the carbon (Fig. 26).

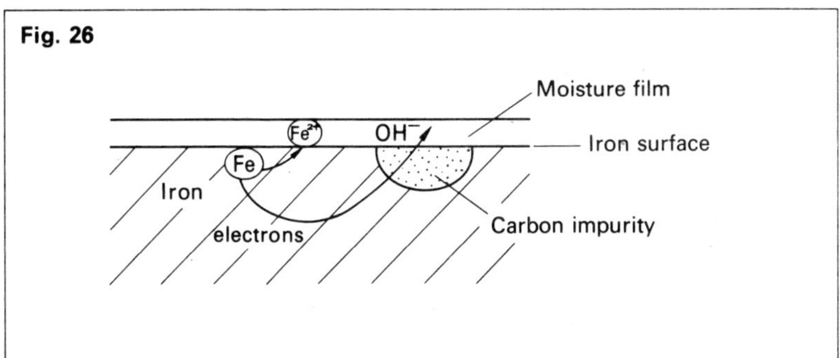

Fig. 26

How the hydroxide ions form when iron rusts

You may wonder how the hydroxide ions are formed during the corrosion process. Experiments indicate that both air and water are required for the corrosion of iron. Since the corrosion of iron is the process of iron forming ions and losing electrons,

$$Fe \longrightarrow Fe^{2+} + 2e$$

then the formation of hydroxide ions must involve the gain of electrons by the water and oxygen of the air.

$$2H_2O + O_2 + 4e \longrightarrow 4OH^-$$

6.8 Protecting iron

Any method where electrons are pushed on to the iron should give good protection from corrosion. Methods used are:–

1 Cathodic protection (Fig. 27)

The iron is connected to the negative terminal of a D.C. power supply. Electrons are pushed on to the iron, preventing it from forming ions.

This can be used to protect steel piers, etc.

Fig. 27

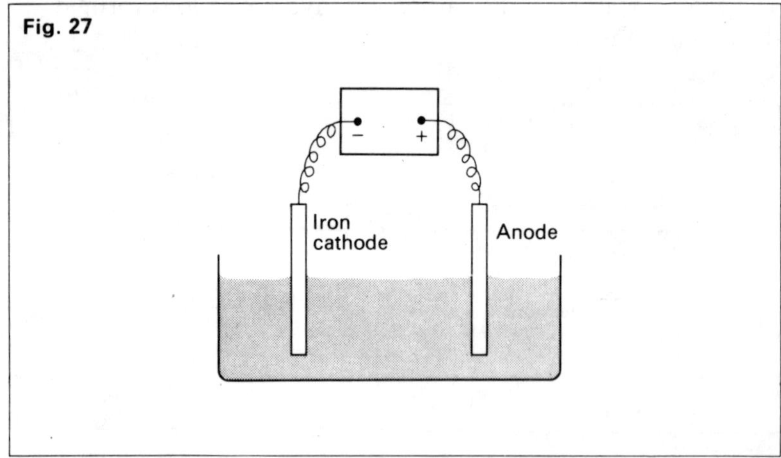

2 Sacrificial protection, using magnesium

Magnesium scrap is connected to the iron, and since it forms ions more readily, electrons are pushed on to the iron, stopping it ionising.

Oil pipe-lines, or steel constructions too expensive to replace, can be protected in this manner.

3 Galvanising (zinc plating)

Fig. 28

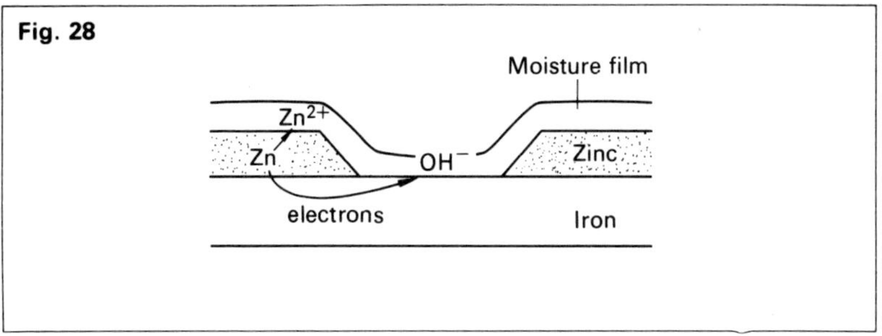

The zinc protects the iron from air and water. When the surface is scratched, and the iron is exposed, the zinc still protects the iron by sacrificial protection (as in Fig. 28).

4 Tin plating (as in 'tin cans')

The tin protects the iron physically. However, when there is a break in the tin, the iron forms ions (corrodes) more rapidly, since tin is less active than iron, and electrons flow from the iron to the tin.

However, tin plating is still used for canning foods because it is much less active than zinc, and is less likely to react with the food if it comes into contact with it.

5 Electroplating

Copper, nickel or chromium are plated on to the iron by making the iron the cathode (negative terminal) of a cell containing the metal salt as electrolyte (Fig. 29).

6 Painting or Greasing

Effect of salt on corrosion rate

Since the process of corrosion involves the setting up of

Fig. 29

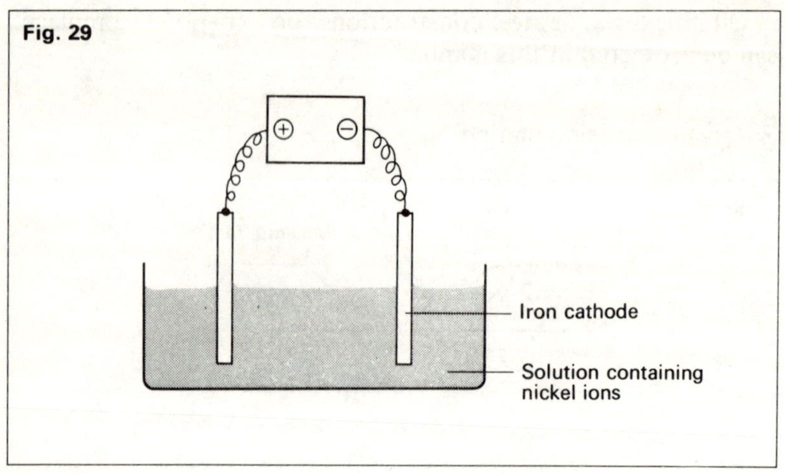

electrolytic cells, an ionic solution (for example, sodium chloride), which is a better conductor than water, will speed up the corrosion rate.

6.9 Aluminium

Aluminium is more reactive than iron, but it does not corrode readily. This is because a thin oxide layer forms on the surface and sticks firmly, protecting the aluminium from further corrosion. (The oxide layer on iron flakes off.)

If aluminium is dipped in mercury (II) chloride solution, some of the mercury ions are displaced, and the mercury atoms formed go on to the aluminium. This does not stop the aluminium oxidising in air, but it does stop the oxide layer sticking to the aluminium surface. When left exposed to air, the aluminium now corrodes very quickly, and flakes of aluminium oxide form.

Anodising aluminium

By this process, the oxide layer on the aluminium is thickened, protecting the aluminium even further from corrosion.

The aluminium is made the *anode* of a cell containing dilute sulphuric acid. When a current is passed through, oxygen is released at the aluminium anode, and thickens the oxide layer on the surface.

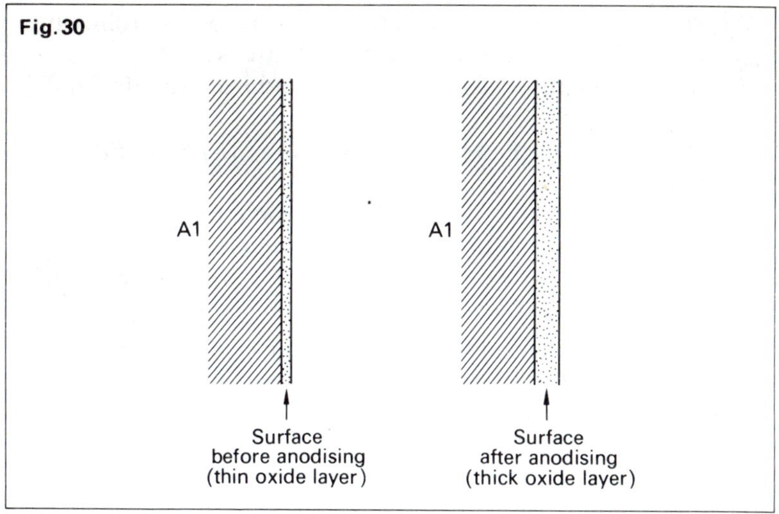

Uses of aluminium

1 The anodised aluminium can be used out of doors without fear of corrosion.
2 Anodised aluminium foil is used in cooking.
3 The thick oxide layer can absorb dyes, and is used for making coloured aluminium pots and pans, and coloured wrappings (e.g. for chocolate biscuits).

The dyeing of aluminium can be demonstrated in the laboratory.

Some revision questions

You should try to answer each question, and then check your answer by referring to the section indicated in brackets after the question.

The electrochemical series is given in the Reduction Potential Tables at the end of this book. You should use them to answer much of the following.

1 What would happen in the following cases? Give the ion-electron equations for any reaction which takes place.

(a) Magnesium added to a solution of copper (II) ions.
(b) Zinc added to a solution of lead ions.
(c) Iron added to a solution of zinc ions. (Sections 6.2 and 6.3.)

2 In which direction would electrons flow when the following pairs of metals are linked up and dipped into water?

(a) Zinc and copper; (b) magnesium and lead; (c) zinc and magnesium. (Section 6.5.)

3 What is meant by the terms (a) oxidation, (b) reduction, (c) redox?

Using the data tables, write the ion electron equations for the reaction between (i) chlorine and iodide ions, (ii) bromine and iron (II) ions. In each case, indicate where oxidation and reduction has taken place. (Section 6.6.)

4 (a) What are the conditions required for iron to corrode?

(b) Describe how galvanising protects iron even once the surface is scratched.

(c) What happens when tinned iron gets scratched?

(d) Aluminium is more active than iron, but it does not corrode readily. Why? (Section 6.7 to 6.9.)

7

Acids and bases

7.1 Acidic and basic oxides

When *non-metals* burn in oxygen, they often form *acidic oxides.* An acidic oxide forms an acidic solution if it dissolves in water (pH paper turns red).

Metals burn to form *basic oxides* (*bases*)

If these basic oxides (bases) dissolve in water, they form *alkaline* solutions (pH paper turns blue).

Many bases, however, do not dissolve in water and do not therefore form alkalis in water (for example, iron oxide and copper oxide).

7.2 Acids and the hydrogen ion

When acid solutions are electrolysed, hydrogen gas is always evolved at the cathode (negative electrode).

Therefore, acid solutions must contain the hydrogen ion, $H^+(aq)$.

What is the hydrogen ion?

Fig. 31

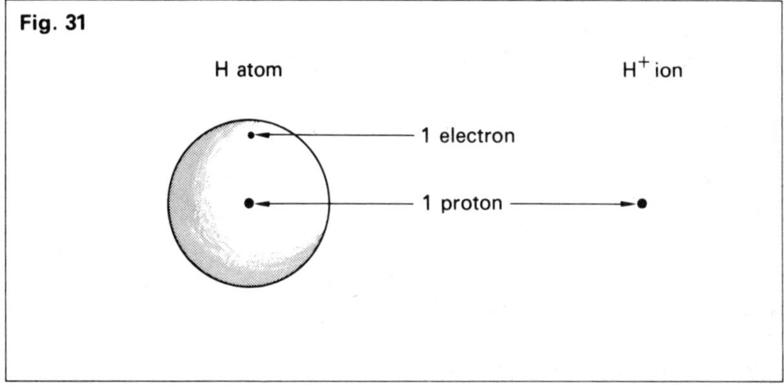

The H^+ ion would simply be a *proton*, with no protective outer electron energy levels.

The unprotected positive charge is very powerful and is automatically attracted to polar water molecules, becoming hydrated (surrounded by water molecules).

$$H^+ + H_2O \longrightarrow H^+(aq)$$

This happens in all acid solutions.

An acid is a hydrogen containing compound, which in solution in water forms hydrated hydrogen ions (H^+ (aq)).

Characteristics of Acids

1 Acids and metals

Dilute acids react with metals from magnesium to tin in the activity series (for safety metals above magnesium are never used with acids).

Hydrogen gas is given off. Examples are:

$$Mg(s) + 2H^+(aq) \longrightarrow Mg^{2+}(aq) + H_2(g)$$
$$Fe(s) + 2H^+(aq) \longrightarrow Fe^{2+}(aq) + H_2(g)$$

Acids + metal oxides (bases)

$$Cu^{2+}O^{2-}(s) + 2H^+(aq) \longrightarrow Cu^{2+}(aq) + H_2O$$
$$Mg^{2+}O^{2-}(s) + 2H^+(aq) \longrightarrow Mg^{2+}(aq) + H_2O$$

The metal ions go into solution, and water is formed.

3 Acid + carbonates

Carbon dioxide gas is released, and water is formed. This happens when the carbonate is present as a solid, or in solution.

$$CO_3^{2-} + 2H^+(aq) \longrightarrow CO_2(g) + H_2O$$

4 Acid + alkali

In each case, water is formed:

$H^+(aq) + OH^-(aq) \longrightarrow H_2O(l)$

7.3 Bases and the hydroxide ion, OH^-

A *base* is usually a metallic oxide or hydroxide. Examples are copper oxide (CuO), sodium hydroxide ($NaOH$) and iron (II) hydroxide ($Fe(OH)_2$).

An *alkali* is the name given to the aqueous solution of a base. All alkalis contain the hydroxide ($OH^-(aq)$) ion.

e.g. $Ca^{2+}O^{2-}(s) + H_2O \longrightarrow Ca^{2+}(aq) + 2OH^-(aq)$
$Na^+OH^-(s) + H_2O \longrightarrow Na^+(aq) + OH^-(aq)$
Base Alkali

Insoluble bases cannot form alkalis. For example, copper oxide and iron (II) hydroxide are both insoluble in water, so no free hydroxide ions can be formed.

7.4 The pH scale

The pH scale is used to define the concentration of hydrogen ions in solution. pH paper, or Universal indicator solution (pH indicator solution) is used to find the pH value of a solution.

pH of acids (which contain $H^+(aq)$ ions)

By repeated dilutions of a molar solution (diluting ten times each dilution) the relationship between pH and concentration of hydrochloric acid can be shown.

Molarity of hydrochloric acid	$\frac{M}{10} = \frac{M}{10^1}$	$\frac{M}{100} = \frac{M}{10^2}$	$\frac{M}{1000} = \frac{M}{10^3}$	$\frac{M}{10\,000} = \frac{M}{10^4}$	$\frac{M}{100\,000} = \frac{M}{10^5}$	$\frac{M}{1\,000\,000} = \frac{M}{10^6}$	$\frac{M}{10\,000\,000} = \frac{M}{10^7}$
pH number	1	2	3	4	5	6	7

The pH of pure water = 7. Therefore, pure water has the same pH as $\frac{M}{10\,000\,000}$ hydrochloric acid. So pure water must contain a very few hydrogen ions.

pH of alkalis (which contain $OH^-(aq)$ ions)

The pH of alkalis varies with concentration in the manner shown below.

Molarity of sodium hydroxide solution	$\frac{M}{10}$	$\frac{M}{100}$	$\frac{M}{1000}$	$\frac{M}{10\,000}$	$\frac{M}{100\,000}$	$\frac{M}{1\,000\,000}$	$\frac{M}{10\,000\,000}$
pH number	13	12	11	10	9	8	7

The pH of pure water = 7 Therefore pure water has the same pH as $\frac{M}{10\,000\,000}$ sodium hydroxide solution.
Pure water must, therefore, contain a very few hydroxide ions. When a water molecule breaks up it must give one hydroxide ion for every hydrogen ion:

$$H_2O(l) \longrightarrow H^+(aq) + OH^-(aq)$$

Thus, at pH 7, the number of hydrogen and hydroxide ions must be the same.

The pH scale

0 1 2 3 4 5 6	7	8 9 10 11 12 13 14
increasing acid strength ←	neutral	increasing base strength (alkali) →

7.5 Strong and weak acids

When the conductivities of $\frac{M}{10}$ hydrochloric acid and $\frac{M}{10}$ ethanoic acid are compared, the hydrochloric acid shows a much higher conductivity.

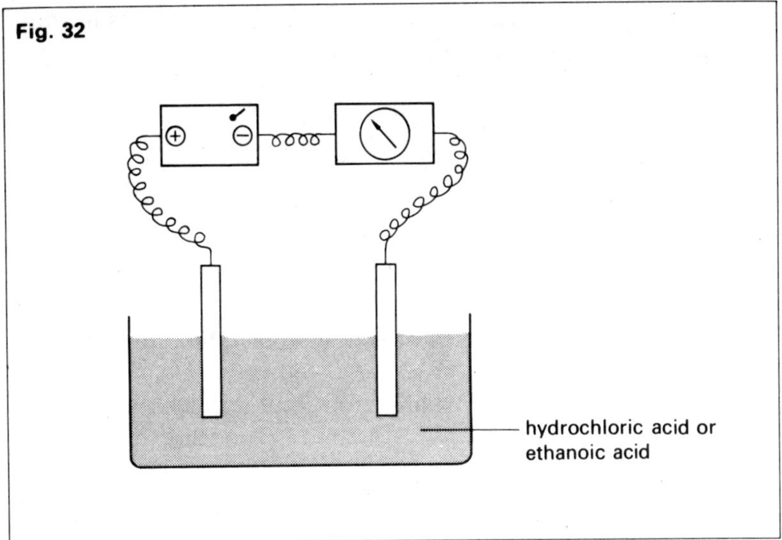

Fig. 32

This indicates that ethanoic acid has fewer ions in solution.

Also, the pH of $\frac{M}{10}$ hydrochloric acid is found to be 1, whereas that of $\frac{M}{10}$ ethanoic acid is found to be 3, again indicating that ethanoic acid has fewer hydrogen ions in solution.

Hydrochloric acid is a *strong acid*. When dissolved in water, it is almost completely dissociated into ions.

$$\underset{\text{hydrogen chloride}}{HCl(g)} + H_2O \longrightarrow \underset{\text{hydrochloric acid}}{H^+(aq) + Cl^-(aq)}$$

Nitric and sulphuric acids are examples of other strong acids.

$$HNO_3(l) + H_2O \longrightarrow H^+(aq) + NO_3^-(aq)$$
$$H_2SO_4(l) + H_2O \longrightarrow 2H^+(aq) + SO_4^{2-}(aq)$$

Ethanoic acid is a *weak acid*. When dissolved in water, it is only dissociated to a very small extent. That is, it exists in solution mainly as undissociated molecules.

$CH_3.COOH + H_2O \rightleftharpoons CH_3COO^-(aq) + H^+(aq)$
mostly in this state — only a few ions present

7.6 Acid strength and concentration

Acid strength is the degree to which an acid will dissociate into ions in aqueous solution.
Concentration is the amount of acid dissolved in a litre of the solution.

Try not to confuse the meaning of these terms.

7.7 Position of bases on the pH scale

7
8
9
10
. . . . lime water (calcium hydroxide)
11
12 . . . . ammonia
13
. . . . sodium hydroxide, potassium hydroxide.
14

Sodium hydroxide is a *strong base* (or strong alkali). That is, it is almost completely dissociated in solution (it is almost completely split up into free sodium ions and free hydroxide ions.

Ammonia solution is a weak base – it only forms a small number of ions in solution.

7.8 The ionisation of water

We originally looked on water as being a polar covalent compound:

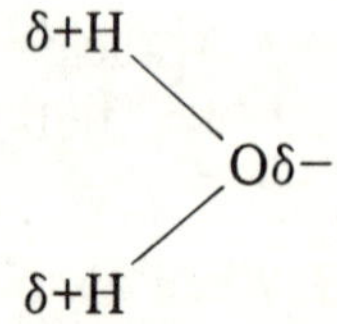

However, experiments show that pure water does conduct electricity slightly. This indicates the existence of a very small number of ions. In addition, the experiments discussed in section 7.4 indicated the existence of a small number of hydrogen and hydroxide ions.

Water molecules dissociate *to a very slight extent* into hydrogen and hydroxide ions:

$H_2O(l)$	$\rightarrow$	$H^+(aq) + OH^-(aq)$
most water is in this form.		only a very small number of ions present.

Since water is a neutral compound (pH = 7), there must be, as shown by the equation, equal numbers of hydrogen and hydroxide ions. (This was also suggested in section 7.4.)

We also know that H^+ ions of an acid and OH^- ions of an alkali will neutralise each other to form water.

$$H^+(aq) + OH^-(aq) \rightarrow H_2O$$

The reaction is therefore *reversible*

$$H_2O(l) \rightleftharpoons H^+(aq) + OH^-(aq)$$

and in water an *equilibrium* is set up in which a few water molecules are always dissociating into ions, while the ions recombine to form water. At any one instant, the proportion of water molecules to hydrogen and hydroxide ions always remains constant.

What happens when hydrogen ions or hydroxide ions are removed from water?

When hydrogen ions are removed from water, the balance is upset, and an excess of hydroxide ions is produced (that is, the solution becomes alkaline). Similarly, when hydroxide ions are removed, an excess of hydrogen ions is produced (that is the solution becomes acidic).

Some revision questions

You should try to answer each question, and then check your answer by referring to the section indicated in brackets after the question.

1 What are acidic and basic oxides? (Section 7.1.)

2 List the main properties of an acid, and give examples of each. (Section 7.2.)

3 What is the connection between a base and an alkali? Why can all alkalis be called bases, but all bases cannot be called alkalis? (Section 7.3.)

4 (a) What is meant by the pH scale?

(b) State the range of pH numbers which would indicate (i) an acid, (ii) an alkali, (iii) a neutral solution.

(c) What is the difference between a *strong* and a *weak* acid? How can a *weak* acid be *concentrated*? (Section 7.4 and 7.5.)

5 Water conducts electricity to a very slight extent. Why? What is meant by saying that water molecules dissociate, and an equilibrium is set up? (Section 7.8.)

8

Neutralization and salt formation

8.1 What is a salt?

Salts all get their names from their parent acids.

For example, sulphuric acid (H_2SO_4) forms sulphates
hydrochloric acid (HCl) forms chlorides
nitric acid (HNO_3) forms nitrates

A salt also contains a positive ion – a positive metal ion or the ammonium, NH_4^+, ion.

We may, therefore, define a salt thus:

A salt is the substance formed when the hydrogen ions of an acid are replaced by metal ions (or the ammonium ion).

8.2 Solubility of ionic compounds

Note: to decide the correct method of preparing a salt, we must know whether various compounds are soluble or insoluble in water. You may find the following table useful.

Compound	Exceptions
Soluble	
1 Ammonium salts	–
2 Salts of Group I metals	–
3 Nitrates	–
4 Sulphates	Lead sulphate and sulphates of calcium, strontium and barium in Group II.
5 Chlorides	Chlorides of lead, silver and mercury (1).
Insoluble	
1 Lead salts	Lead nitrate
2 Carbonates	Ammonium carbonate, carbonates of Group I metals.
3 Oxides and hydroxides	Group I and Group II metal oxides or hydroxides.

8.3 Preparation of salts

(a) Soluble salts prepared by neutralization

1 *Acid + active metal* (down to tin in the Activity Series). Take the preparation of magnesium sulphate as an example.

$$Mg(s) + H_2SO_4(aq) \longrightarrow MgSO_4(aq) + H_2(g)$$

Crystals of magnesium sulphate are obtained by concentrating the solution (boiling off most of the water), and leaving to crystallise.

The general ionic equation is:

$$Mg(s) + 2H^+(aq) \longrightarrow Mg^{2+}(aq) + H_2(g)$$

2 *Acid + basic oxide.* Take the preparation of copper (II) sulphate as an example.

$$CuO(s) + H_2SO_4(aq) \rightarrow CuSO_4(aq) + H_2O$$

Blue copper sulphate crystals are obtained by concentrating the solution.
Ionic equation: $Cu^{2+}O^{2-}(s) + 2H^+(aq) \longrightarrow Cu^{2+}(aq) + H_2O$

3 *Acid + alkali.* Take the preparation of sodium chloride as an example.

$$NaOH(aq) + HCl(aq) \rightarrow NaCl(aq) + H_2O$$

Ionic equation: $OH^-(aq) + H^+(aq) \rightarrow H_2O$

4 *Acid + carbonate.* An alternative preparation of copper sulphate provides an example.

$$H_2SO_4(aq) + CuCO_3(s) \rightarrow CuSO_4(aq) + H_2O + CO_2(g)$$

Ionic equation: $Cu^{2+}CO_3^{2-}(s) + 2H^+(aq) \rightarrow Cu^{2+}(aq) + H_2O + CO_2(g)$

Carbon dioxide gas is evolved, and the copper sulphate crystals are obtained from the solution as before.

(b) Insoluble salts – prepared by precipitation

If a salt is insoluble it is prepared by adding a solution of the positive metal ions to a solution of the negative ions.

Take the preparation of barium sulphate as an example.

$$Ba^{2+}(aq) + SO_4^{2-}(aq) \longrightarrow Ba^{2+}SO_4^{2-}(s)$$

The barium and sulphate ions may come from a variety of compounds which contain them.

e.g. $BaCl_2(aq) + Na_2SO_4(aq) \longrightarrow BaSO_4(s) + 2NaCl(aq)$

soluble barium chloride; soluble; insoluble

or $Ba(OH)_2(aq) + H_2SO_4(aq) \longrightarrow BaSO_4(s) + 2H_2O$

soluble

Whenever a barium ion and a sulphate ion come together. water is unable to separate them again, and an insoluble precipitate is formed.

Only in one of the neutralization reactions indicated above is it obvious that the hydrogen ions are being removed from the acid, since hydrogen gas is liberated. In the other cases, we have only *assumed* that water is formed in using up the hydrogen ions. We must try to find some experimental evidence to prove this.

8.4 Mobility of ions – another approach to neutralization

Earlier experiments on conductivity (2.1, 2.2) may have shown that not all electrolytes have equal conductivities.

Further experiments can suggest possible explanations.

(a) Concentration effect

In general we find that the *higher the concentration of the ions, the higher the conductivity.*

(b) Mobility (speed) of ions

To investigate effects other than concentration, we must use different solutions of equal concentration.

Results obtained from experiments on some of these solutions would be:

(i) hydrochloric acid ($H^+ + Cl^-$): best conductor
(ii) sodium hydroxide ($Na^+ + OH^-$)
(iii) sodium chloride ($Na^+ + Cl^-$): poorest conductor.

Comparing the results of (i) and (iii), the chloride ion (Cl^-) is common to both, and any differences must be due to the hydrogen (H^+) or sodium (Na^+) ions.

Therefore, the hydrogen ion (H^+) is faster than the sodium ion (Na^+).

In the same way, comparing (ii) and (iii), the Na^+ ion is common, therefore the hydroxide ion (OH^-) is faster than the chloride ion (Cl^-).

(c) Comparing the speeds of hydrogen ions and hydroxide ions

If an experiment is set up in which an indicator shows the movement of hydrogen ions towards the cathode, and of the hydroxide ions towards the anode, then the hydrogen ion is seen to be about twice as fast as the hydroxide ion.

Fig. 33 Movement of H^+ and OH^- ions

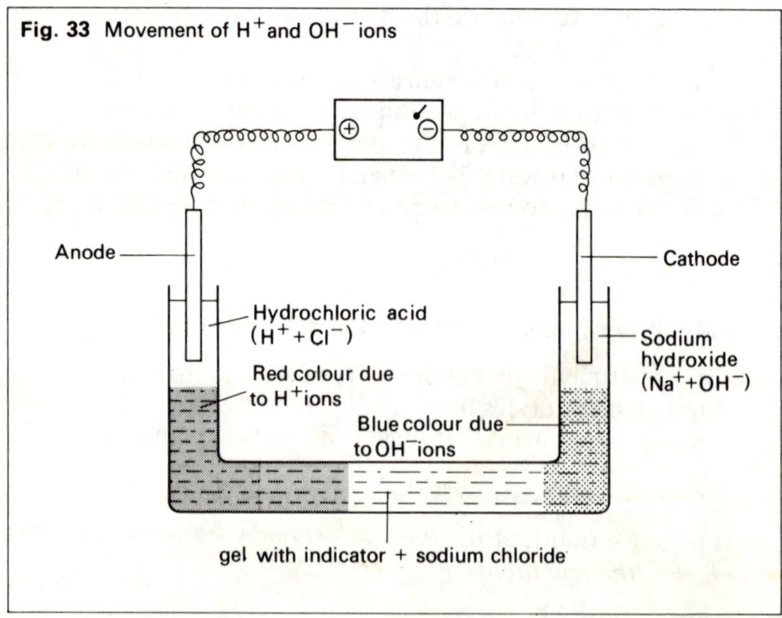

To sum up: hydrogen ions are faster than hydroxide ions which are faster than all the others.

Therefore, if hydrogen ions or hydroxide ions are lost from a solution (perhaps being replaced by slower moving ions) then the conductivity of the solution should decrease.

The experiments on salt preparations could now be repeated, using conductivity measurements.

8.5 Magnesium added to dilute hydrochloric acid

The graph in Fig. 34 is obtained. The reaction is:

$$2H^+Cl^-(aq) + Mg(s) \longrightarrow Mg^{2+}(Cl^-)_2(aq) + H_2(g)$$

Fig. 34

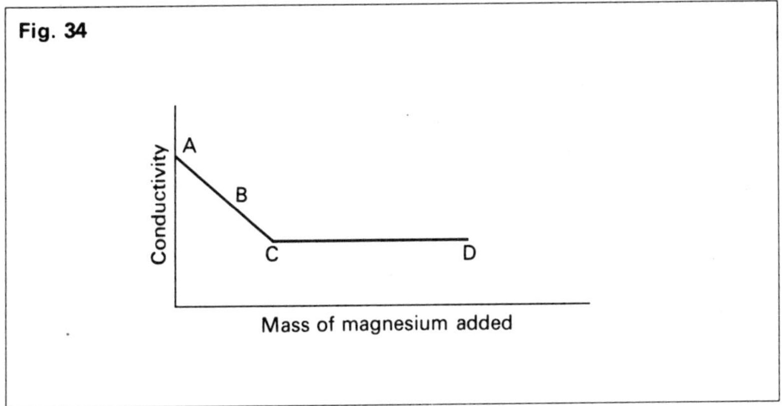

Explanation of graph

	Ions in solution
At A No magnesium has been added, so there are only $H^+(aq)$ and $Cl^-(aq)$ ions present. H^+ ions are fast. Therefore, the conductivity is high.	H^+ Cl^- H^+ Cl^- H^+ Cl^- H^+ Cl^-
At B Some magnesium has been added. This reacts, and replaces some of the H^+ ions with Mg^{2+} ions. The $H^+(aq)$ ions form hydrogen gas. Therefore, the current drops.	Mg^{2+} Cl^- Cl^- H^+ Cl^- H^+ Cl^-
At C Sufficient magnesium has now been added to neutralize the acid, and all the fast moving H^+ ions have been replaced by slower moving Mg^{2+} ions. Therefore, the current reaches its lowest point.	Mg^{2+} Cl^- Cl^- Mg^{2+} Cl^- Cl^-
From C to D The conductivity remains steady, since there are no more H^+ ions to be replaced, and the magnesium has only a very slow reaction with water.	no change Mg^{2+} Cl^- Cl^- Mg^{2+} Cl^- Cl^-

Since Cl^-(aq) ions are unchanged throughout, the reaction can be written:

$$Mg(s) + 2H^+(aq) \rightarrow Mg^{2+}(aq) + H_2(g)$$

8.6 Acid + alkali (sodium hydroxide solution added to hydrochloric acid)

The graph in Fig. 35 is obtained. The reaction is:

$$Na^+OH^-(aq) + H^+Cl^-(aq) \rightarrow Na^+Cl^-(aq) + H_2O$$

The reaction can also be followed by adding methyl orange indicator to the acid before the reaction.

Fig. 35

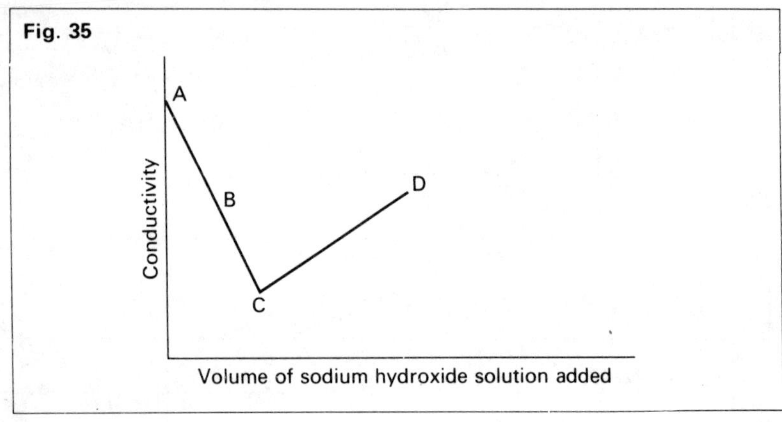

Explanation of Graph

	Ions in solution	
At A H^+ and Cl^- ions present.	H^+	Cl^-
At B Half the NaOH required has been added. H^+(aq) replaced by Na^+(aq), due to $H^+(aq) + OH^-(aq) \rightarrow H_2O$. Therefore, the conductivity drops.	Na^+ H^+	Cl^- Cl^-
At C The ***end point*** (that is, reaction is complete). All the H^+ ions have now been replaced, so the conductivity is at its lowest. The indicator changes colour at this point, showing that neutralization is complete.	Na^+ Na^+	Cl^- Cl^-
C to D Excess NaOH being added. No H^+ left. OH^- ions now present (fast moving), so the conductivity rises.	Na^+ Na^+ Na^+	Cl^- Cl^- OH^-

Note: Since the Na^+(aq) and Cl^-(aq) ions are spectator ions, the

$$H^+(aq) + OH^-(aq) \longrightarrow H_2O(l)$$

8.7 Acid + carbonate (solid copper (II) carbonate added to sulphuric acid solution)

The graph in Fig. 36 is obtained. The reaction is
reaction may be written

$$(H^+)_2SO_4{}^{2-}(aq) + Cu^{2+}CO_3{}^{2-}(s) \rightarrow Cu^{2+}SO_4{}^{2-}(aq) + H_2O + CO_2(g)$$

Fig. 36

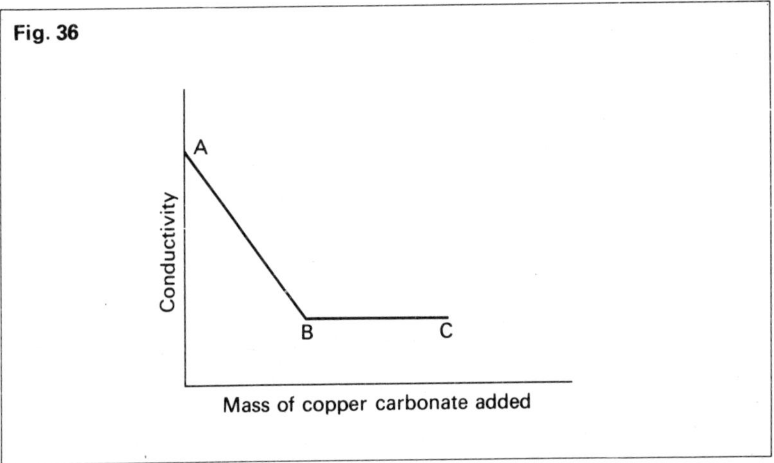

Explanation of graph

	Ions in solution	
At A High conductivity, due to H^+ ions	H^+ H^+	$SO_4{}^{2-}$
At B All the H^+ ions have been replaced, and the only ions in solution are Cu^{2+} and $SO_4{}^{2-}$	Cu^{2+}	$SO_4{}^{2-}$
B to C As excess copper carbonate is added, there is no further change in conductivity since it is insoluble in water.	Cu^{2+}	$SO_4{}^{2-}$

The $SO_4{}^{2-}$ ions are spectator ions, so the reaction is

$$2H^+(aq) + Cu^{2+}CO_3{}^{2-}(s) \rightarrow Cu^{2+}(aq) + H_2O + CO_2(g)$$

The copper ions cannot be called spectator ions since they start as $Cu^{2+}(s)$ and finish as $Cu^{2+}(aq)$.

8.8 Precipitation (barium hydroxide solution added to dilute sulphuric acid)

The graph in Fig. 37 is obtained. The reaction is:

$$(H^+)_2SO_4{}^{2-}(aq) + Ba^{2+}(OH^-)_2(aq) \longrightarrow Ba^{2+}SO_4^{2-}(s) + 2H_2O$$

Fig. 37

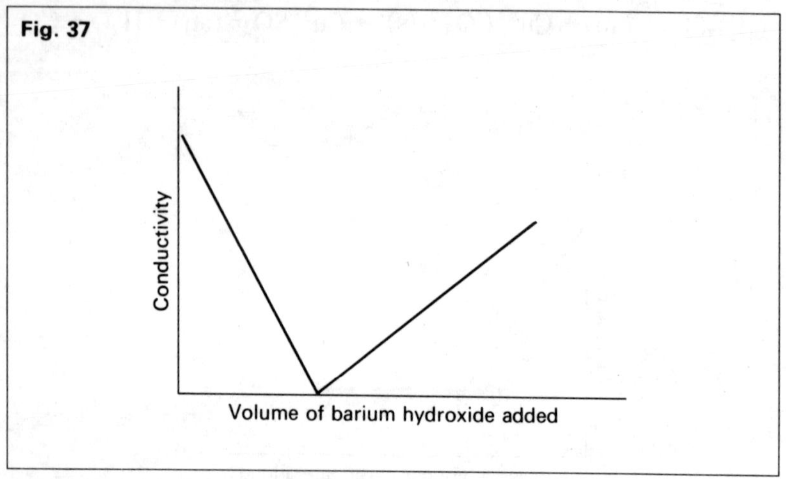

The conductivity is high at the start, due to the presence of the hydrogen ions of the acid.

As barium hydroxide solution is added, two reactions are taking place:

$H^+(aq) + OH^-(aq) \rightarrow H_2O$ (H^+ ions removed from solution)
(being added)

$SO_4{}^{2-}(aq) + Ba^{2+}(aq) \rightarrow Ba^{2+}SO_4{}^{2-}(s)$
(being added)
insoluble, so $SO_4{}^{2-}$ ions are being removed from solution.

Therefore, at the *end point*, there are no ions left in solution, and the conductivity is zero.

After the end point, barium hydroxide solution is being added and the conductivity rises steeply, due to the mobility of the hydroxide ions.

Note: Since water shows very slight conductivity, the conductivity would not quite reach zero if a sensitive meter was used.

8.9 Heats of neutralization

Compare the reactions:

(a) $H^+Cl^-(aq) + Na^+OH^-(aq) \rightarrow Na^+Cl^-(aq) + H_2O$

Omitting spectator ions (Na^+ and Cl^- ions):

$$H^+(aq) + OH^-(aq) \rightarrow H_2O$$

(b) $(H^+)_2SO_4^{2-}(aq) + 2Na^+OH^-(aq) \rightarrow (Na^+)_2SO_4^{2-}(aq) + 2H_2O$

Omitting spectator ions (Na^+ and SO_4^{2-} ions)

$$2H^+(aq) + 2OH^-(aq) \rightarrow 2H_2O$$

or $H^+(aq) + OH^-(aq) \rightarrow H_2O$

That is, in acid/alkali neutralizations, the only energy change is due to the formation of water. So if equal numbers of hydrogen ($H^+(aq)$) and hydroxide ($OH^-(aq)$) ions are mixed, the heat change should be the same. This can be checked by experiment.

If the experiment was carried out using sulphuric acid solution and barium hydroxide solution, the rise in temperature will be higher than expected.

This is because there are two reactions taking place.

1 $H^+(aq) + OH^-(aq) \rightarrow H_2O$
2 $Ba^{2+}(aq) + SO_4^{2-}(aq) \rightarrow Ba^{2+}SO_4^{2-}(s)$

The second reaction accounts for the additional heat evolved.

8.10 Determining the concentration of a solution

These experiments are usually carried out using indicators, and not by conductivity experiments.

Examples

1 25 cm^3 of M hydrochloric acid neutralized 50 cm^3 of sodium hydroxide solution. Find the molarity of the sodium hydroxide.
Balanced equation:

$$HCl(aq) + NaOH(aq) \rightarrow NaCl(aq) + H_2O$$

From equation: 1 mole of HC1 reacts with 1 mole of NaOH
i.e. 1 litre of M HCl reacts with 1 litre of M NaOH
∴ 25 cm^3 of M HCl reacts with 25 cm^3 of M NaOH
In other words, if the NaOH solution was molar, the volume used would have been 25 cm^3.

But, in the experiment, 50 cm^3 NaOH solution was used.
∴ The NaOH must have been spread through twice the amount of water.

∴ The concentration of the sodium hydroxide solution $= \frac{M}{2}$

2 100 cm^3 of M potassium hydroxide (KOH) was required to neutralize 25 cm^3 of sulphuric acid solution. What was the concentration of the acid?
Balanced equation:

$$2KOH(aq) + H_2SO_4(aq) \rightarrow K_2SO_4(aq) + 2H_2O$$
2 moles 1 mole

From equation: 2 moles KOH reacts with 1 mole H_2SO_4
i.e. 2 litres of M KOH reacts with 1 litre of M H_2SO_4
∴ 100 cm^3 of M KOH reacts with 50 cm^3 of M H_2SO_4
In other words, if the sulphuric acid was $\frac{M}{1}$, the volume used would be 50 cm^3.
But, in the experiment, 25 cm^3 of acid was used,
i.e. The required amount of acid was contained in only half the volume of solution, so it must have been twice as concentrated.
∴ Concentration of sulphuric acid = 2M.

Some revision questions

You should try to answer each question, and then check your

answer by referring to the section indicated in brackets after the question.

1 Suggest how you might prepare samples of the following salts: (a) magnesium sulphate, (b) sodium nitrate, (c) copper carbonate. (Section 8.2 and 8.3.)

2 In an experiment you are asked to measure the conductivity of dilute sulphuric acid as sodium hydroxide solution is added to it.

Write the equation for the reaction, and draw and explain the resulting conductivity graph you would expect.
(Section 8.4 and 8.6.)

3(a) 25 cm^3 of $\frac{M}{1}$ nitric acid neutralized 25 cm^3 of sodium hydroxide solution. Fine the molarity of the sodium hydroxide.

(b) 50 cm^3 of $\frac{M}{1}$ sodium hydroxide neutralized 50 cm^3 of sulphuric acid. Find the molarity of the acid
(Section 8.10)

9

Electrolysis of aqueous solutions

9.1 Copper (II) chloride solution

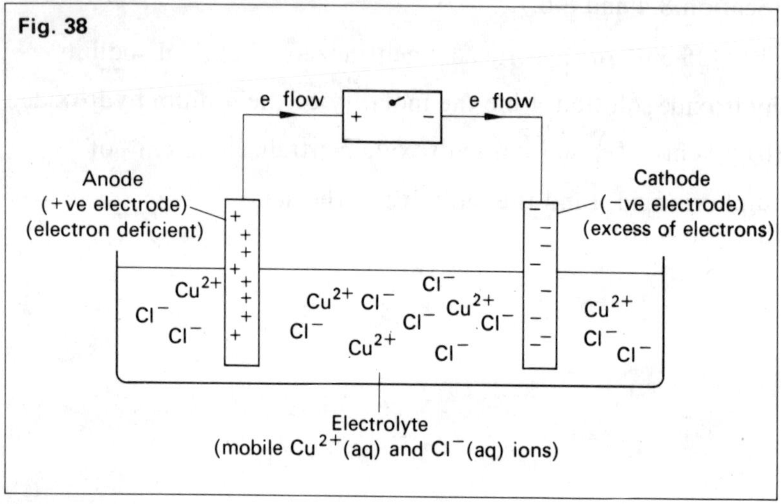

Fig. 38

At Anode (Fig. 39)

Negative chloride (Cl^-) ions are attracted to the anode. Since the anode is short of electrons, it pulls electrons from the chloride ions, and chlorine gas (Cl_2) is discharged.

$Cl^- \rightarrow Cl + e$
then $Cl + Cl \rightarrow Cl_2$
(*or* $2Cl^- \rightarrow Cl_2 + 2e$)

At Cathode (Fig. 40)

Copper ions collect.
The cathode has an excess of electrons, so it pushes off 2e onto each copper ion, and copper metal is formed.

$Cu^{2+} + 2e \rightarrow Cu$

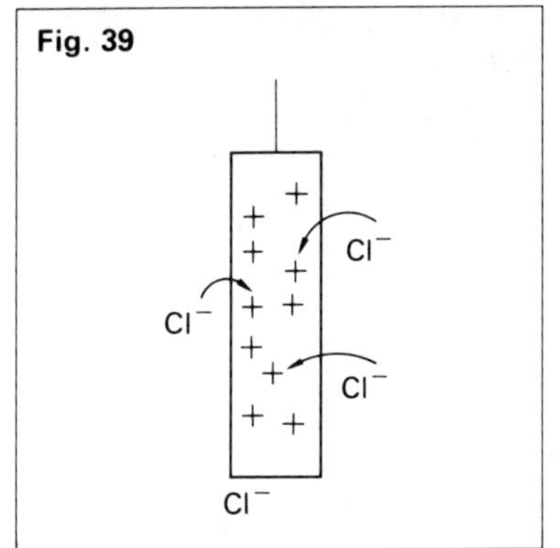

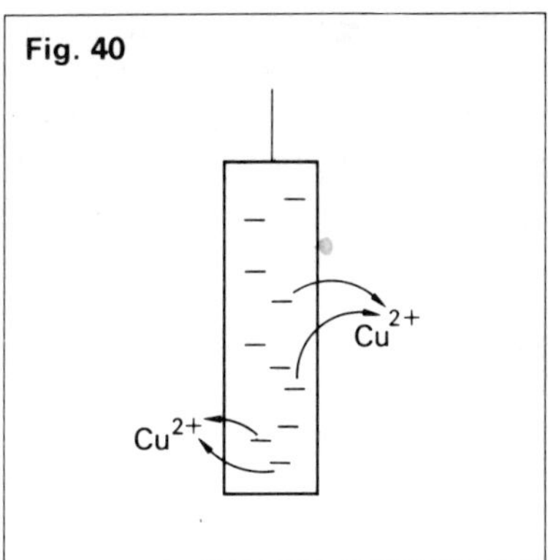

9.2 Electrolysis of sodium halides

The table below shows the major products formed.

	Salt	Cathode(–)	Anode(+)
(a)	Sodium fluoride	Hydrogen	Oxygen
(b)	Sodium chloride	Hydrogen	Chlorine
(c)	Sodium bromide	Hydrogen	Bromine
(d)	Sodium iodide	Hydrogen	Iodine

We showed in sections 7.4 and 7.8 that water contains a small number of ions:

$$H_2O(l) \rightleftharpoons H^+(aq) + OH^-(aq)$$

Therefore, in any solution, these ions must also be available for discharge.

We can use this fact to help explain the results of the electrolysis.

Cathode Reaction

In all cases, hydrogen gas is formed at the cathode, and *not* sodium.

The hydrogen ($H^+(aq)$) ions from the water accept electrons more readily than the sodium ions ($Na^+(aq)$).

Therefore hydrogen ions from the water are discharged as hydrogen gas:

$$2H^+(aq) + 2e \rightarrow H_2(g)$$

When the hydrogen ions are removed from the water, there will be an excess of hydroxide ($OH^-(aq)$) ions.

The solution at the cathode becomes alkaline. This can be shown by an indicator solution.

Anode Reaction

(*b*) $2Cl^-(aq) \rightarrow Cl_2 + 2e^-$
(*c*) $2Br^-(aq) \rightarrow Br_2 + 2e^-$
(*d*) $2I^-(aq) \rightarrow I_2 + 2e^-$

In (*a*) the fluoride ion will not readily lose electrons to form fluorine, because fluorine has a very strong electron attracting power. Therefore, the fluoride ion is *not* discharged. Instead, hydroxide ions from the water discharge as oxygen gas.

$$4OH^-(aq) \rightarrow 2H_2O + O_2(g) + 4e$$

Since hydroxide ions from the water are discharged, an excess of hydrogen ions build up. Therefore the solution at the anode becomes acidic (shown by the addition of an indicator).

9.3 Electrolysis of dilute sulphuric acid ('electrolysis of water')

Cathode	*Anode*
$H^+(aq)$ ions collect	$SO_4^{2-}(aq)$ and $OH^-(aq)$ from water collect.
$H^+(aq)$ ions discharge	$OH^-(aq)$ ions from the water discharge in preference to the SO_4^{2-} ions.
$2H^+(aq) + 2e \rightarrow H_2(g)$	$4OH^-(aq) \rightarrow 2H_2O + O_2(g) + 4e$

Let us balance the reactions at the anode and cathode, so that equal numbers of electrons are transferred.

$4H^+(aq) + 4e \rightarrow 2H_2$	$4OH^-(aq) \rightarrow 2H_2O + O_2 + 4e$

For every four hydrogen ions discharged at the cathode, four hydroxide ions are discharged at the anode.

The volume of hydrogen released at the cathode is twice the volume of oxygen released at the anode.

In other words, water is being removed, and the electrolysis is equivalent to the electrolysis of water.

9.4 Preferential order of discharge of ions

Cathode reaction

The order below can be deduced from the data tables.

$Ag^+(aq) + e \rightarrow Ag$ most easily discharged
$Cu^{2+}(aq) + 2e \longrightarrow Cu$
$2H^+(aq) + 2e \longrightarrow H_2$

Ca^{2+}, Na^+, Li^+, K^+ etc., are *never* discharged from aqueous solution. H^+ ions from the water are discharged in preference.

Anode reaction

The order for chloride, bromide and iodide can be deduced from the data tables.

$2I^-(aq) \longrightarrow I_2 + 2e$ most easily discharged
$2Br^-(aq) \longrightarrow Br_2 + 2e$
$2Cl^-(aq) \longrightarrow Cl_2 + 2e$
$4OH^-(aq) \longrightarrow 2H_2O + O_2 + 4e^-$

$F^-(aq)$, $SO_4^{2-}(aq)$, $NO_3^-(aq)$ are *never* discharged from aqueous solution. OH^- ions from the water are discharged in preference.

Note. The order of discharge shown here is satisfactory for our purposes. However, if you continue with chemistry, you will find that the order will alter slightly.

Some revision questions

Copy and complete the following table, writing equations for the reactions at the electrodes, of the solutions listed below.

Solution of	Reaction at cathode	Reaction at anode
1 Copper (II) bromide, $CuBr_2$		
2 Silver nitrate, $AgNO_3$		
3 Sodium sulphate, Na_2SO_4		
4 Potassium iodide, KI		
5 Lithium fluoride LiF		
6 Copper (II) sulphate $CuSO_4$		

10

Sulphur and its compounds

10.1 Sulphuric acid

Sulphuric acid is one of the most widely used chemicals in any industrial society. Therefore, manufacture of this acid is of vital importance to the nation's economy.

Production of sulphuric acid must be based on the raw materials available. The main sources are:

1 The element sulphur – found uncombined in Texas and Sicily;

2 Iron pyrites ore (FeS_2) and other sulphide ores.

Both sulphur and iron pyrites can yield the gas sulphur dioxide. We will pause to consider this gas before continuing with sulphuric acid.

10.2 Sulphur dioxide (SO_2)

Tests for sulphur dioxide

(a) It turns *starch iodate* paper or solution a deep blue colour. This is a very sensitive test.

(b) It decolourises yellow-brown bromine water.

(c) It decolourises dark brown iodine solution.

Other materials react in the same way to tests (b) and (c), so these tests are not conclusive.

Obtaining sulphur dioxide from the raw materials

(a) Burning sulphur in air

$$\underset{\text{(from air)}}{S + O_2} \longrightarrow SO_2$$

(b) Heating iron pyrites, or any other sulphide ore, in air. Iron oxide and sulphur dioxide are formed.

$$4FeS_2 + 11O_2 \rightarrow 2Fe_2O_3 + 8SO_2$$

Properties of sulphur dioxide

1 It has a very unpleasant smell and irritates the throat and lungs.
2 It is easily liquefied – a mixture of solid carbon dioxide and alcohol will produce a temperature of about −70°C. This liquefies the sulphur dioxide.

It is therefore stored in cylinders as a liquid under pressure, rather than as a compressed gas.
3 It is very soluble in water. An acid solution (sulphurous acid) is formed.

$$H_2O + SO_2 \rightarrow \underbrace{2H^+(aq) + \overset{\text{sulphite ion}}{SO_3^{2-}(aq)}}_{\text{sulphurous acid}}$$

4 Although the gas does not burn, and does not usually support combustion (that is, other materials will not burn in sulphur dioxide), burning magnesium will continue to burn, forming sulphur and magnesium oxide.

$$2Mg + SO_2 \rightarrow 2MgO + S$$

Sulphur dioxide and pollution

Air pollution is a problem in all industrial areas, and one of the major pollutants is sulphur dioxide.

Sulphur or sulphur compounds are found in many fuels – for example, heavy fuel oils, coal, etc. Therefore, any industry which burns these fuels produces sulphur dioxide. Coal or oil-burning power stations, town gas production and the burning of town gas itself, and the steel industry all produce quantities of sulphur dioxide. The extremely poisonous nature of the gas makes it vital that the amount of sulphur dioxide is kept down to a tolerable level.

10.3 Reactions of the sulphite ion, SO_3^{2-}

In order to carry out experiments involving the sulphite ion, we must first be able to detect the presence of the *sulphate* ion, SO_4^{2-}.

Test for the sulphate ion in solution

If hydrochloric acid is added to the suspected sulphate, and then a solution containing barium (Ba^{2+}) ions is added (usually barium chloride), a white precipitate is produced if the sulphate ion is present.

$$SO_4^{2-}(aq) + Ba^{2+}(aq) \rightarrow \underset{\text{white precipitate}}{Ba^{2+}SO_4^{2-}(s)}$$

The sulphite ion is a reducing agent (electron donor)

When reacting with materials which can be made to accept electrons, the sulphite ion readily donates electrons, and in so doing is itself oxidised to the sulphate ion.

$$SO_3^{2-} + H_2O \rightarrow SO_4^{2-} + 2H^+ + 2e$$

(This equation can be found in the 'Standard Reduction Electrode Potential' tables at the end of this book.)

The formation of the sulphate ion can be proved in the experiments below, by carrying out the test for the sulphate ion as detailed above.

(a) *With the halogens (Br_2 Cl_2 I_2)*

On adding bromine water to a solution containing sulphite ions, the red-brown bromine water is decolourised.

The bromine molecules have been reduced (gained electrons) to bromide ions.

$$\underset{\text{brown}}{Br_2} + 2e \rightarrow \underset{\text{colourless}}{2Br^-}$$

similarly,

$$\underset{\text{green}}{Cl_2} + 2e \rightarrow \underset{\text{colourless}}{2Cl^-}$$

$$\underset{\text{brown}}{I_2} + 2e \rightarrow \underset{\text{colourless}}{2I^-}$$

(b) *With iron (III) ions (Fe^{3+})*

The colour of the solution after mixing with sulphite ion solution changes from yellow to green, showing that Fe^{3+} ions have been reduced to Fe^{2+} ions.

$$\underset{\text{yellow}}{Fe^{3+}} + e \rightarrow \underset{\text{green}}{Fe^{2+}}$$

Note: All the above reduction equations can be found in the standard tables.

Proof of electron transfer in the above reactions (Fig. 41)
The solution in the right hand limb can be bromine water, iron (III) ion solution, silver ion solution, or any other electron acceptor.

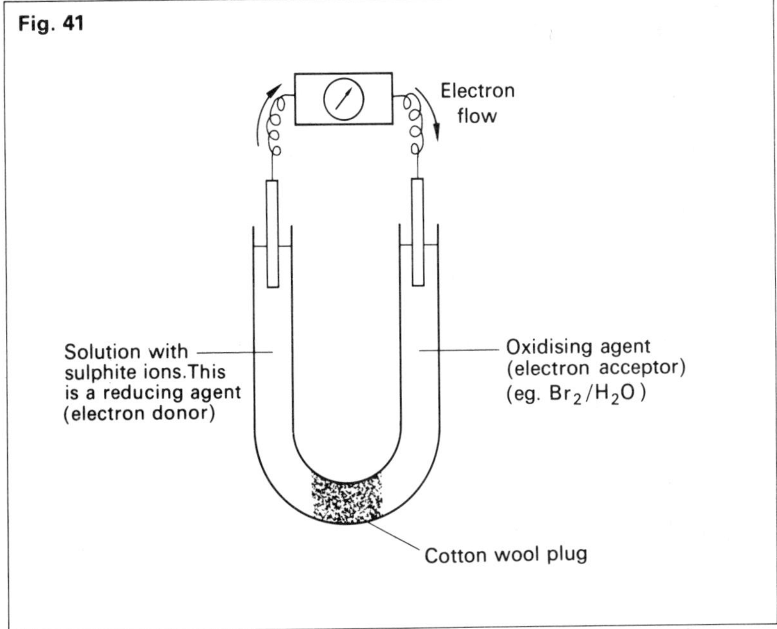

Electrons flow from the sulphite ion solution to the bromine water (shown by the meter).

10.4 Uses of sulphites

(a) *Bleaching.* The sulphite ion can bleach straw, and can safely bleach woollens and other delicate fabrics. This is again a reducing action, and when left exposed to air and sunlight paper is re-oxidised and turns yellow again.
(b) As a *disinfectant* and *preservative.* Due to its highly poisonous nature, trace amounts of sulphur dioxide can kill bacteria and micro-organisms, and it can therefore be used to preserve wines and foods.
(c) As a useful *source of sulphur dioxide.* Addition of moderately concentrated hydrochloric or sulphuric acid to any metal sulphite, and warming where necessary, yields sulphur dioxide gas.

$$SO_3^{2-} + 2H^+ \rightarrow H_2O + SO_2$$

For example: $Na_2SO_3 + 2HCl \rightarrow 2NaCl + H_2O + SO_2$

10.5 Sulphur trioxide (SO_3) and sulphuric acid (H_2SO_4)

The preparation of pure sulphuric acid by oxidation of the sulphite ion is too difficult to carry out on a large scale. Instead, sulphur dioxide is made to combine directly with oxygen, forming sulphur trioxide, which can be converted to sulphuric acid by dissolving in water.

The reaction: $2SO_2 + O_2 \rightarrow 2SO_3$

involves the collision of sulphur dioxide and oxygen molecules of sufficient energy to cause the change to take place.

At room temperature the energy of collision is not great enough for any noticeable reaction to take place. Therefore we must increase the speed of the molecules by raising the temperature.

However, when sulphur trioxide is heated, it breaks down to sulphur dioxide and oxygen.

$$2SO_3 \xrightarrow{\text{heat}} 2SO_2 + O_2$$

In other words, the reaction is reversible:

$$2SO_2 + O_2 \rightleftharpoons 2SO_3$$

and at any temperature an *equilibrium* is reached where sulphur trioxide molecules are being broken down as fast as they are being made, and the percentage of sulphur trioxide formed depends on the temperature. In this case, the lower the temperature the higher the percentage of sulphur trioxide.

However, the lower the temperature, the lower the energy of collision of the molecules, and hence the longer it takes to reach equilibrium.

Our problem is:

High temperature – reaction fast, but % SO_3 formed too small.

Low temperature – % SO_3 high, but reaction too slow.

To speed up the rate at the lower temperature we use a catalyst (in this case *vanadium* (v) *oxide*, or *platinum* spread over asbestos wool), which provides a big surface area for the molecules to react on. A temperature of about 450°C has been found to be a reasonable temperature for this reaction.

Preparation of sulphur trioxide and sulphuric acid

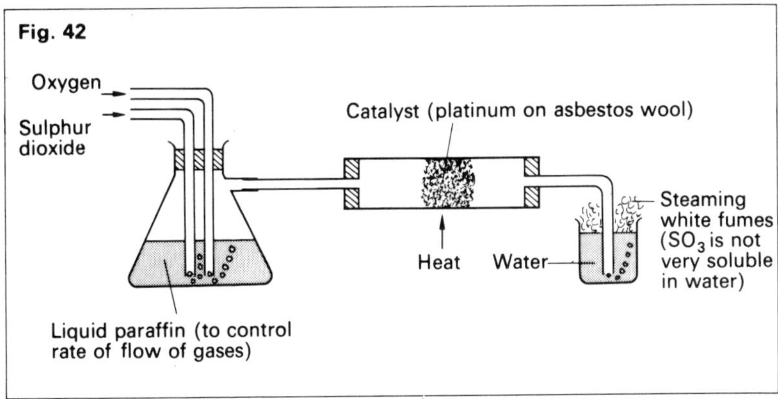

$$2SO_2 + O_2 \xrightarrow[\text{heat (450°C)}]{\text{catalyst (platinum)}} 2SO_3$$

If the sulphur trioxide obtained is dissolved in water, sulphuric acid is obtained:

$$SO_3 + H_2O \rightarrow H_2SO_4$$

This process is important in industry, and is called the Contact Process.

The sulphur trioxide is not very soluble in water, and in industry it is normally dissolved in some concentrated sulphuric acid first, forming *oleum*, and the correct amount of water is then added to form sulphuric acid.

10.6 Industrial preparation of sulphuric acid – The Contact Process

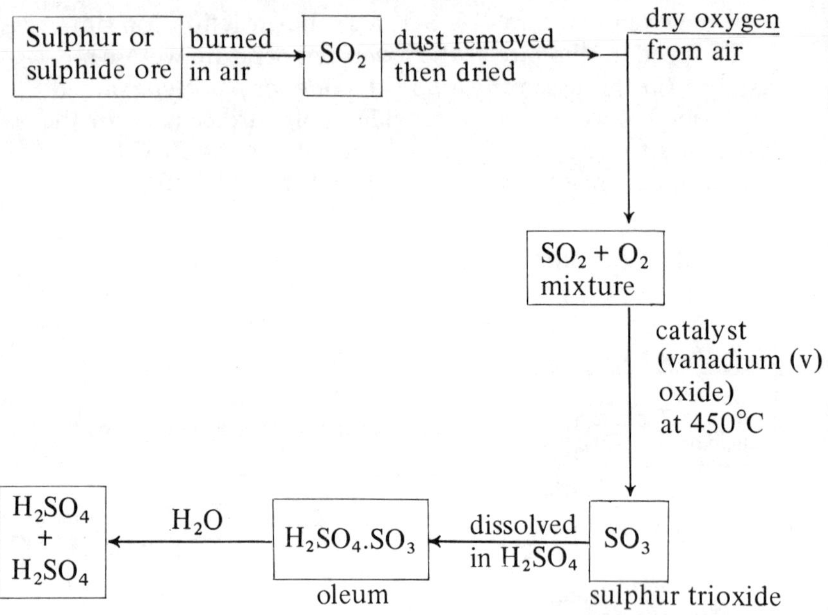

10.7 Properties of dilute sulphuric acid

This is a strong acid which is fully ionised when in solution in water. The ions are:

$2H^+(aq) + SO_4^{2-}(aq)$

It has all the properties associated with a strong acid:

(a) pH number is very low;

(b) it forms carbon dioxide gas when added to carbonates:

$CO_3^{2-}(s) + 2H^+(aq) \rightarrow H_2O + CO_2(g)$
metal
carbonate
(e.g. Na_2CO_3)

(c) its hydrogen ions can be easily replaced:

(i) with an active metal

$$Mg(s) + 2H^+(aq) \rightarrow Mg^{2+}(aq) + H_2(g)$$

(ii) by adding a base

$$H^+(aq) + OH^-(aq) \rightarrow H_2O$$

e.g. $H_2SO_4 + 2NaOH \rightarrow Na_2SO_4 + 2H_2O$

or $2H^+(aq) + O^{2-}(s) \rightarrow H_2O$

e.g. $H_2SO_4 + CuO \rightarrow CuSO_4 + H_2O$

10.8 Properties of concentrated sulphuric acid

1 The acid has a high boiling point, and this property helps us to prepare other acids with lower boiling points. For example, in the preparation of hydrochloric acid, sulphuric acid is added to a chloride.

Fig. 43

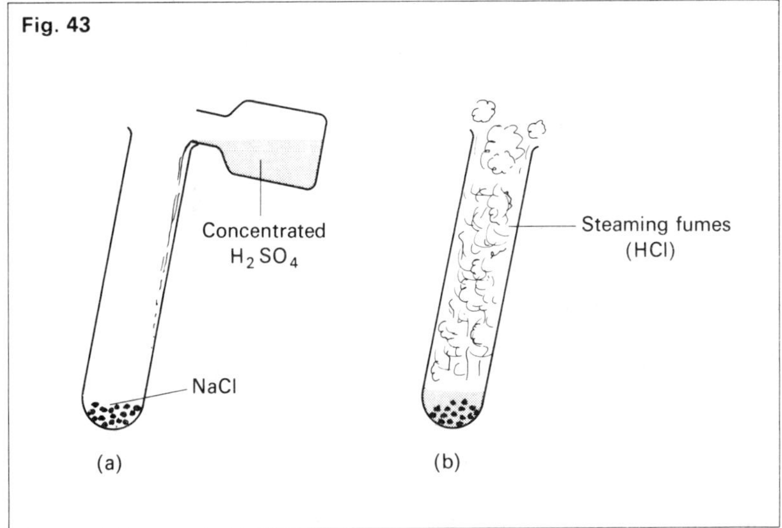

$$H_2SO_4(l) + Cl^- \longrightarrow HSO_4^- + HCl(g)$$

(as NaCl) (hydrogen sulphate ion)

On dissolving the hydrogen chloride gas in water, hydrochloric acid is formed.

In the case of nitric acid, sulphuric acid is added to a nitrate:

$$H_2SO_4 + NO_3^- \longrightarrow HSO_4^- + HNO_3$$

2 It has a great attraction for water, and will absorb water from the air. It is therefore a good drying agent for gases which do not react with it.

The affinity of sulphuric acid for water is so great that it will remove the elements of water from compounds which contain them:

(a) blue hydrated copper sulphate crystals turn white (become anhydrous) – the water of crystallisation is removed;

(b) sugar (sucrose) chars, and swells up as a black mass of carbon:

$$C_{12}H_{22}O_{11} \rightarrow 12C + 11H_2O$$

Fig. 44

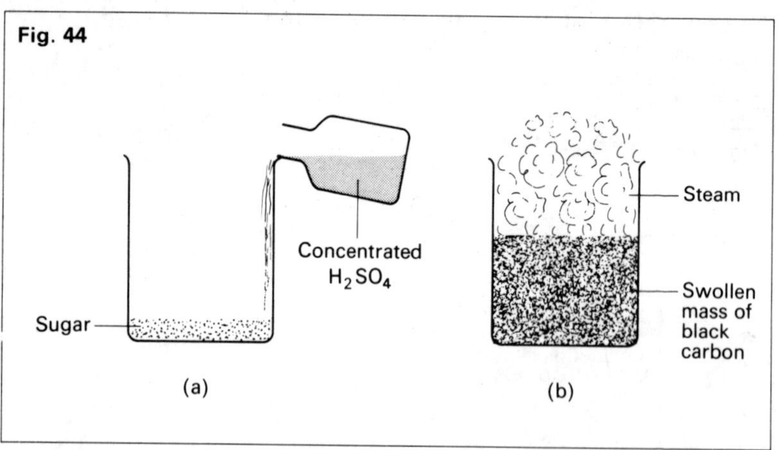

(c) paper, wood, etc., becomes charred as the water is removed. The attraction of concentrated sulphuric acid for water is so great, and so much heat is given out, that the acid should *always* be added to water and *never* water added to the acid. This is because the heat produced could turn drops of water added into steam, causing acid to spurt out.

Concentrated Sulphuric acid is NOT highly ionised

(a) Blue litmus paper, one half wet, lies on top of some concentrated sulphuric acid. The wet half turns red immediately, the dry part stays blue.
(b) Pure, concentrated sulphuric acid is a poor conductor of electricity. It therefore contains only a few ions.

Concentrated sulphuric acid contains only a few ions, but when added to water it dissociates, forming a lot of $H^+(aq)$ ions. It is therefore a strong acid.

$$H_2SO_4(l) + H_2O(l) \rightarrow 2H^+(aq) + SO_4^{2-}(aq)$$

4. *Concentrated sulphuric acid is an oxidising agent* (electron acceptor).
(a) *Reaction with metals.* When hot, sulphur dioxide gas (SO_2) is formed. With copper and zinc.

$$Cu \longrightarrow Cu^{2+} + 2e$$
$$Zn \longrightarrow Zn^{2+} + 2e$$

In the above cases, the sulphuric acid accepts electrons, and is reduced to sulphur dioxide gas.

Note: whereas copper will not react with dilute acids, it will react with concentrated sulphuric acid as above.

(b) *Reaction with non-metals.* With carbon the reaction is:

$$C + H_2SO_4 \rightarrow CO_2 + SO_2 + H_2O$$

Once again, the acid must be hot.

10.9 Uses of sulphuric acid

(a) In steel pickling – dipping steel into acid to clean it before treatment (for example before chrome plating).
(b) Manufacture of fertilisers – superphosphates and ammonium sulphate.
(c) Dyeing – many organic dyestuffs are made using sulphuric acid.
(d) Manufacture of artificial fibres and plastics.
(e) Manufacture of soaps and detergents.

(f) Manufacture of other acids, e.g. hydrofluoric acid (HF) – used in glass etching.

Some revision questions

You should try to answer each question and then check your answer by referring to the section indicated in brackets after the question.

1 What acid forms when sulphur dioxide dissolves in water? Write its formula. (Section 10.2.)

2 (a) Give the test for the sulphate ion in solution.
(b) What is meant when it is said that the sulphite ion is a reducing agent? Give two examples of its reducing action. You should use the Reduction Potential Tables to help you write any equations. (Section 10.3.)

3 Give the conditions necessary and state the reactions involved in the preparation of sulphuric acid from sulphur dioxide. (Section 10.5.)

4 (a) How can sulphuric acid be used to prepare other acids? Give an example.
(b) State some ways in which the properties of dilute and concentrated sulphuric acid differ.
(Sections 10.7 and 10.8.)

1. Sulphurous acid. H_2SO_4

2.
a) Take the soln containing the sulphate add hydrochloric acid and then add a soln containing Barium ions ; a white precipitate is produced is the Sulphate is present

b) An electron donor.

$Br_2 + 2e \longrightarrow 2Br^-$

$Cl_2 + 2e \longrightarrow 2Cl^-$
(green) (colourless)

3.
Low temperature. Fast reaction

$SO_2 + O_2 \xrightarrow[\text{heated}]{\text{catalyst}} SO_3 + H_2O \longrightarrow H_2SO_4$

11

Nitrogen and nitrogen compounds

11.1 Nitrogen and nitric acid

Nitrogen is the unreactive gas which occupies 4/5 of the air by volume.

The route to nitric acid formation can be similar to the route for sulphuric acid formation:

$$N_2 \xrightarrow{O_2} \text{Oxides of nitrogen} \xrightarrow{H_2O} \underset{\text{(nitric acid)}}{HNO_3}$$

Nitrogen does not react readily with oxygen, but it can be made to combine if enough energy is used.

Fig. 45

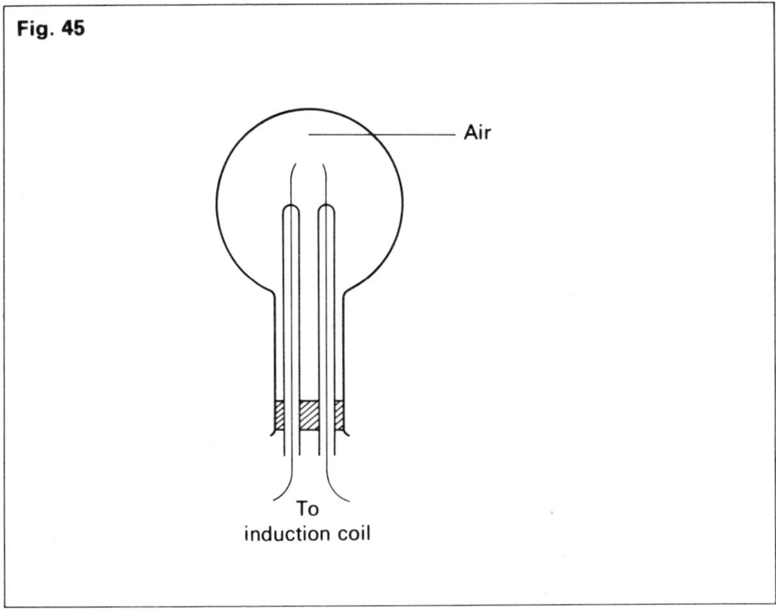

Using an induction coil, sparks are passed through the oxygen/ nitrogen (air) mixture in the flask (as in Fig. 45).

The brown gas, nitrogen dioxide, (NO_2) appears.

When water is added, pH paper turns red – nitric acid has been formed.

This method is *not economic, and so cannot be used.*

A similar reaction to the one just described takes place in air during lightning storms.

11.2 Nitrogen and hydrogen – ammonia (NH_3)

There are many *ammonium compounds* in the laboratory, e.g. ammonium chloride (NH_4Cl), and we can obtain ammonia gas from any ammonium compound by heating it with an alkali, e.g. sodium hydroxide, or calcium hydroxide.

NH_4^+	$+ OH^-$	$\rightarrow NH_3$	$+ H_2O$
e.g. from ammonium chloride	from sodium hydroxide	ammonia	

The gas has a distinctive smell (it should be smelled *carefully*). Moist litmus or pH paper turns blue (hydroxide ions, OH^- are formed).

11.3 Ammonia and water – ammonia solution is a weak base

Ammonia is very soluble in water and forms an alkaline solution.

$$NH_3(g) + H_2O \rightarrow NH_4^+(aq) + OH^-(aq)$$

alkaline due to presence of hydroxide ions.

However, if the conductivity of equal volumes of molar sodium hydroxide and molar ammonia solutions are compared, it is seen that the conductivity of the sodium hydroxide is high, while that of the ammonia solution is low.

Since both the sodium ions and the ammonium ions have nearly the same mobility and OH^- ions are common to both, the large difference in conductivity can only be explained by there being fewer ions present in the ammonia solution.

Sodium hydroxide is a *strong base.* When dissolved in water, it separates into free sodium ions and hydroxide ions.

$Na^+OH^-(s) + H_2O \rightarrow Na^+(aq) + OH^-(aq)$

An ammonia solution of the same concentration is a *weak base* and contains only a few ions. That is, at any one instant only a small number of ammonia molecules have combined with water molecules to form ions:

$NH_3 + H_2O \rightleftharpoons$	$NH_4^+(aq) + OH^-(aq)$
most of the solution is in this form	only a few ions free at any instant

11.4 Combining nitrogen and hydrogen to form ammonia

In trying to form ammonia, we can try different energy sources.

Heating a mixture of nitrogen and hydrogen, or *sparking* the mixture does *not* yield ammonia.

In fact, if the experiment is tried in reverse, it is found that sparking dry ammonia will decompose it into the elements nitrogen and hydrogen (Fig. 46):

$2NH_3 \rightarrow N_2 + 3H_2$

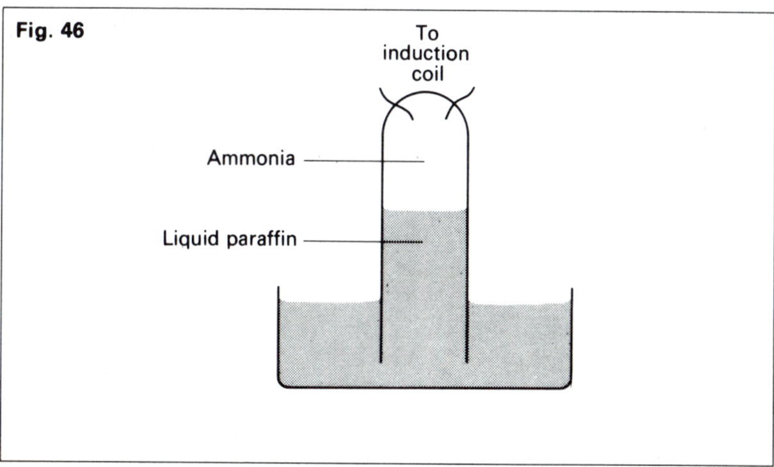

Fig. 46

The Haber Process

The reaction between nitrogen and hydrogen is reversible:

$$N_2 + 3H_2 \rightleftharpoons 2NH_3$$

Depending on conditions, the reaction will be driven towards the right or the left.

The reaction involves the collision of molecules of sufficient energy for the change to take place.

Effect of temperature

Since heating causes the ammonia to break down, then the temperature must be kept as low as possible.

But, if the temperature is too low, the reaction becomes very slow.

In other words:

High temperature	–	reaction fast, but % NH_3 formed is too small.
Low temperature	–	% NH_3 is high, but reaction too slow.

Effect of catalyst

To speed up the rate at the lower temperature, we use a catalyst (in this case, iron oxide), which provides a large surface area for the molecules to react on.

Effect of pressure

If we look at the gas volumes involved:

$$\underbrace{N_2 + 3H_2} \rightleftharpoons 2NH_3$$

$$\text{4 molecules} \rightleftharpoons \text{2 molecules}$$

The 2 molecules of ammonia take up less room than the 4 molecules of mixture.

In other words, in forming ammonia we are trying to convert the gas into a smaller volume. Therefore, increasing the pressure should encourage the formation of ammonia.

A suitable balance of all three conditions must be found, and in practice, the conditions used are:

Temperature	–	about 500°C
Pressure	–	about 1000 atmospheres
Catalyst	–	iron or iron oxide.

Although the percentage of ammonia produced is not high, the process is economical since any uncombined nitrogen and hydrogen can be recycled after removal of the ammonia.

11.5 Oxidation of ammonia

Since the manufacture of nitric acid by oxidation of nitrogen is not economical, we need to find a method of oxidising the ammonia to the oxides of nitrogen, and hence to nitric acid.

Methods which might be attempted are:

1 Burning ammonia in oxygen

Ammonia will not burn readily in air, but it will do so in oxygen, the products being water and nitrogen.

Therefore, this is of *no value* in forming the oxides of nitrogen.

Fig. 47

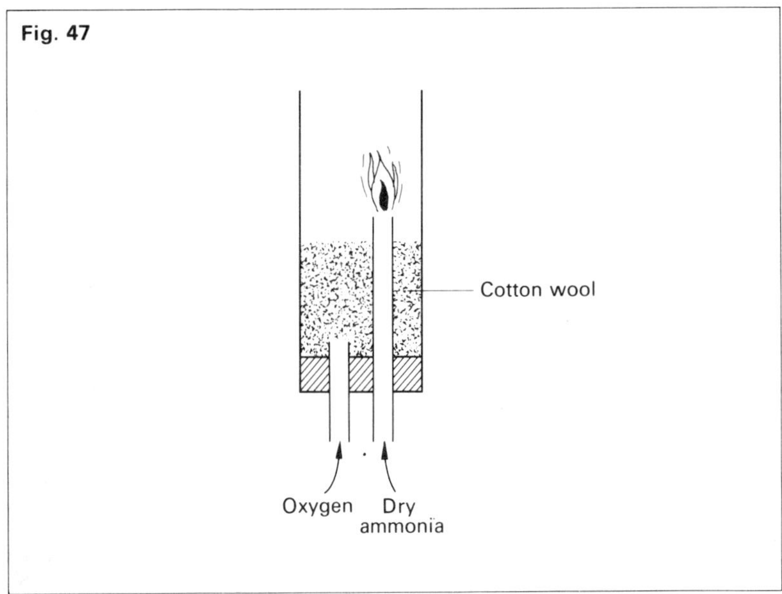

2 Supplying oxygen from another compound

Ammonia is passed over hot copper (II) oxide.
Once again, nitrogen gas and water are formed.

$$NH_3(g) + CuO(s) \rightarrow N_2(g) + Cu(s) + H_2O(l)$$

3 Using a catalyst – The Ostwald Process

If oxygen is bubbled through ammonia solution, and a hot platinum wire lowered into the gas mixture, a reaction occurs, which is sometimes explosive. The reaction must be between the ammonia and oxygen, since oxygen bubbling through water shows no reaction.

Fig. 48

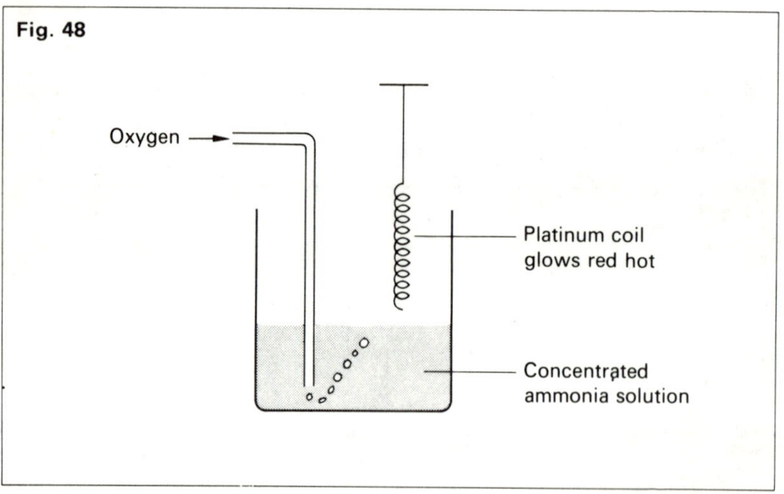

The explosive reaction of the oxygen/ammonia mixture can be controlled if air is used instead of oxygen. The air is bubbled through ammonia solution, and the mixture of gases passed over a platinum catalyst. Excess ammonia can be removed, and brown fumes of nitrogen dioxide collect. When moist pH paper is placed in the gas, it turns red. That is, the solution of the gas is acidic, and is nitric acid.

The main reactions taking place are shown below. For simplicity, they are unbalanced.

a $NH_3 + O_2 \xrightarrow[\text{catalyst}]{\text{hot platinum}} NO + H_2O$
nitrogen monoxide (nitric oxide) (colourless gas)

b $NO + O_2 \longrightarrow NO_2$
(O_2 from air in the receiver) nitrogen dioxide (brown gas)

c $NO_2 + H_2O \longrightarrow HNO_3$
nitric acid

This oxidation is used in industry, and is called the Ostwald Process.

11.6 Test for the nitrate ion, NO_3^-

(a) *Brown ring test*
Iron (II) sulphate solution is added to a solution of the suspected nitrate, and then a slow stream of concentrated sulphuric acid is poured down the side. This forms a layer at the bottom of the test-tube, and at the junction of the two layers a brown ring is obtained if the nitrate ion is present.

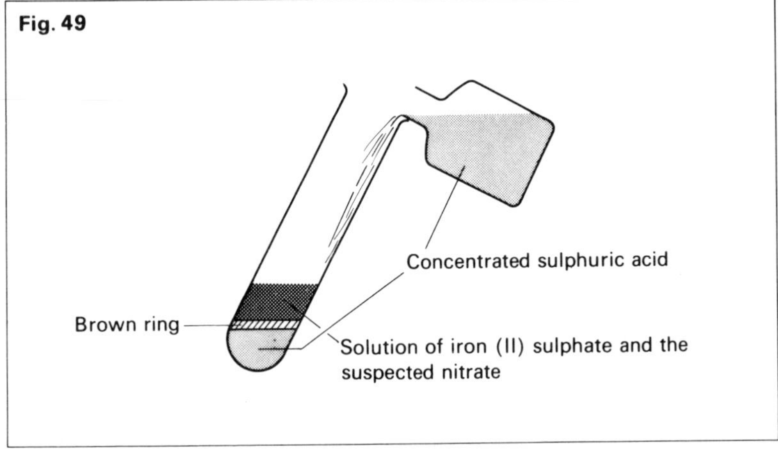

Fig. 49

(b) If copper and concentrated sulphuric acid is added to a nitrate (solid or solution), brown nitrogen dioxide gas is obtained.

11.7 Properties of nitric acid

Dilute nitric acid shows normal acid reactions

1 It turns pH paper red. It is a strong acid, and dissociates almost completely.

$$HNO_3(l) \xrightarrow{H_2O} H^+(aq) + NO_3^-(aq)$$

2 It is neutralized by an alkali (e.g. sodium hydroxide solution) or a base (e.g. copper oxide).

For example: $HNO_3(aq) + NaOH(aq) \rightarrow H_2O + NaNO_3(aq)$
or: $H^+(aq) + OH^-(aq) \rightarrow H_2O$

3 It reacts with carbonates, yielding carbon dioxide and water.

e.g.: $CuCO_3 + 2HNO_3 \rightarrow Cu(NO_3)_2 + H_2O + CO_2$
or: $CO_3^{2-} + 2H^+(aq) \rightarrow H_2O + CO_2(g)$

The metal nitrate is left in solution.

4 The action of nitric acid with metals is different from normal (see below).

11.8 Nitric Acid is an electron acceptor (an oxidising agent)

(i) Reaction of very dilute nitric acid with magnesium (only) (Fig. 50(a))
Hydrogen is evolved:

$Mg(s) \rightarrow Mg^{2+}(aq) + 2e$
$2H^+(aq) + 2\ e \rightarrow H_2(g)$

Fig. 50

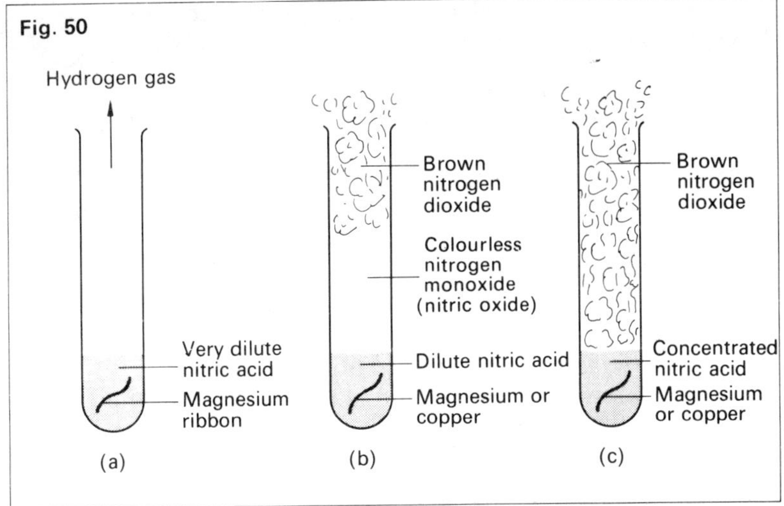

(ii) Reaction of dilute nitric acid on magnesium or copper (Fig. 50(b)).

The magnesium or copper loses electrons (is oxidised).

$$Mg(s) \rightarrow Mg^{2+}(aq) + 2e \quad \textit{or} \quad Cu(s) \rightarrow Cu^{2+}(aq) + 2e$$

The nitrate ion accepts these electrons and forms nitrogen monoxide (nitric oxide) gas (the nitrate ion is reduced).

$$\underbrace{NO_3^-(aq) + 4H^+(aq)}_{\text{from the acid}} + 3e \rightarrow \underset{\substack{\text{nitrogen}\\ \text{monoxide}\\ \text{(nitric oxide)}}}{NO} + 2H_2O$$

When the nitrogen monoxide comes into contact with the air at the top of the tube, it reacts to form brown nitrogen dioxide gas.

(iii) Reaction of concentrated nitric acid with magnesium or copper (Fig. 50(c))
As before, the magnesium or copper loses electrons (is oxidised):

$$Mg(s) \rightarrow Mg^{2+}(aq) + 2e \quad \textit{or} \quad Cu(s) \rightarrow Cu^{2+}(aq) + 2e$$

But this time, the nitrate ion accepts the electrons and forms brown nitrogen dioxide gas:

$$\underbrace{NO_3^-(aq) + 2H^+(aq)}_{\text{from acid}} + e \rightarrow NO_2(g) + H_2O$$

Note: All the ion-electron equations given here can be obtained from the Reduction Electrode Potential Tables at the end of this book.

Note: The gas which we have called nitrogen dioxide and given the formula NO_2, is more correctly called dinitrogen tetroxide, with formula N_2O_4.
As a result the equation for the reduction of the nitrate ion will now read:

$$2NO_3^-(aq) + 4H^+(aq) + 2e \rightarrow N_2O_4(g) + H_2O$$

This is the equation which you will find in the Reduction Electrode Potential Tables.

11.9 The importance of nitrogen compounds

Nitrogenous fertilisers

Plants and animals need to form nitrogen compounds (as proteins). Since nitrogen gas is very unreactive, it must be obtained in the form of nitrogen compounds.

Animals get the nitrogen by eating other animals or plants. Plants must obtain their nitrogen from the soil. We can help the plants to do this by adding fertilisers.

Fertilisers used are:

ammonium sulphate	–	$(NH_4)_2SO_4$
ammonium nitrate	–	NH_4NO_3
ammonium hydrogen phosphate	–	$(NH_4)_2HPO_4$

The Nitrogen Cycle

A few plants such as clover, bean and pea can use atmospheric nitrogen to build protein. These plants have *nodules* on the roots which contain *nitrifying bacteria*, which convert atmospheric nitrogen into nitrates.

If all the nitrogen was put back into the soil as animal manure, compost, etc., and there was rotation of crops, there would be no need to use fertilisers. However, with most food consumed in cities, well away from the centre of growth, the Nitrogen Cycle is broken.

So the Nitrogen Cycle is 'assisted' by 'fixing' nitrogen by the Haber Process, and making fertilisers.

Fig. 51

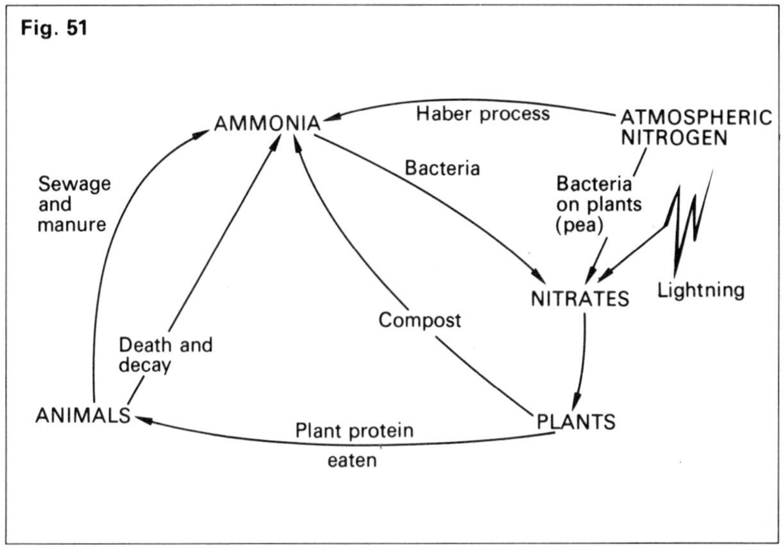

The production of ammonia requires large amounts of fuel, and unless new fuel reserves are found, they cannot last for ever, so our problem has not been solved once and for all.

If the soil is acidic, denitrifying bacteria convert the nitrates back into nitrogen.

Lime can be used to reduce the acidity of the soil.

11.10 Conservation of sewage

Another important element for plant and animal growth is phosphorus.

Animal bones are composed of calcium phosphate, which is insoluble in water. Therefore, application of powdered bone

(bone meal) or crushed calcium phosphate rocks is very slow acting.

Soluble phosphates are:

(a) ammonium hydrogen phosphate;

(b) superphosphate (phosphate rocks treated with sulphuric acid form this fertiliser).

However, the bulk of phosphates and much of the nitrogen is lost to the sea as sewage, and since supplies of phosphate rocks are not unlimited, treatment of sewage and dustbin refuse gives an excellent and cheap compost, rich in nitrogen and phosphorus compounds.

Some Scottish counties have led the way in this process.

Some revision questions

You should try to answer each question and then check your answer by referring to the section indicated in brackets after the question.

1 Suggest any method by which you could quickly prepare a sample of ammonia in the laboratory. (Section 11.2.)

2 What is meant by saying that ammonia is a weak base? (Section 11.3.)

3 Write the equation for the reaction and give the main conditions for the formation of ammonia from its elements. What is the name given to this industrial process? (Section 11.4.)

4 Write the equations, and give the main conditions for the Ostwald Process for the oxidation of ammonia to nitric acid. (The equations need not be balanced.) (Section 11.5.)

5 List the main properties of dilute nitric acid, and indicate where they differ from other dilute acids. (Sections 11.7 and 11.8.)

6 Write down what you can about the nitrogen cycle, and the importance of nitrogen in everyday life. (Section 11.9.)

12

Fuels and related substances

12.1 The element carbon

Of the 100 or so elements on the Periodic Table, the element carbon is found in about $\frac{2}{3}$ of all known compounds. Carbon compounds are usually called *organic compounds*, and because they are so many and so varied they are studied in a separate branch of chemistry called organic chemistry.

First of all, let us look at the element carbon, a constituent in all these compounds.

12.2 Polymorphs of carbon

Carbon can exist in two different crystalline forms, having different physical properties. We call these forms *polymorphs* of carbon. **Polymorphs are different crystalline forms of the same element.**

The two polymorphs are: (a) diamond; (b) graphite.

Whereas diamond forms hard, clear, octahedral (eight-sided) crystals, and is a non-conductor of electricity, graphite forms soft plates and is a good conductor.

Explanation

In *diamond*, each carbon atom is linked covalently to four other carbon atoms, all through the crystal (Fig. 52) and a *macromolecule* is formed:

A macromolecule is a giant molecule, often consisting of more than 2000 atoms.

In *graphite*, each carbon atom has only three strong bonds. The fourth half filled orbital on each carbon only forms loose bonds with surrounding carbon atoms. The electrons in these orbitals are, therefore, fairly free, and graphite is a conductor.

Graphite only forming three strong bonds leads to the formation of flat plates (Fig. 53).

Fig. 52 Part of a diamond

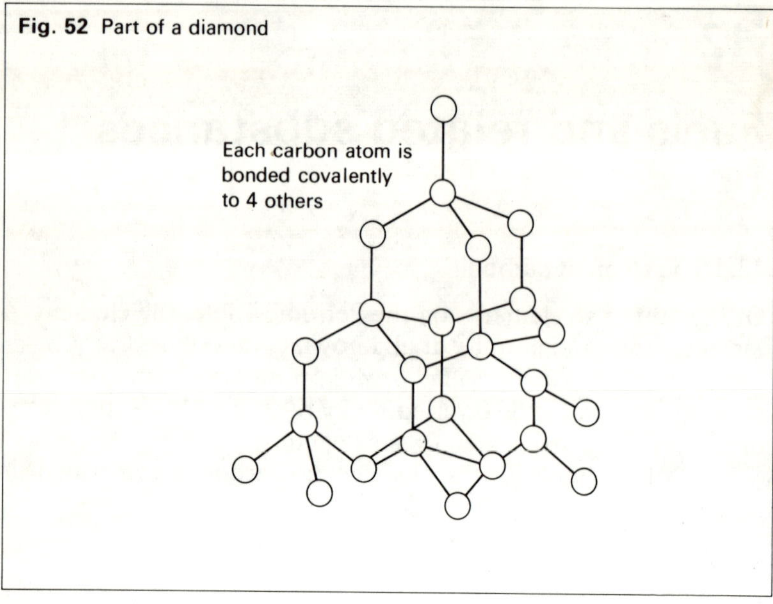

Fig. 53 Graphite

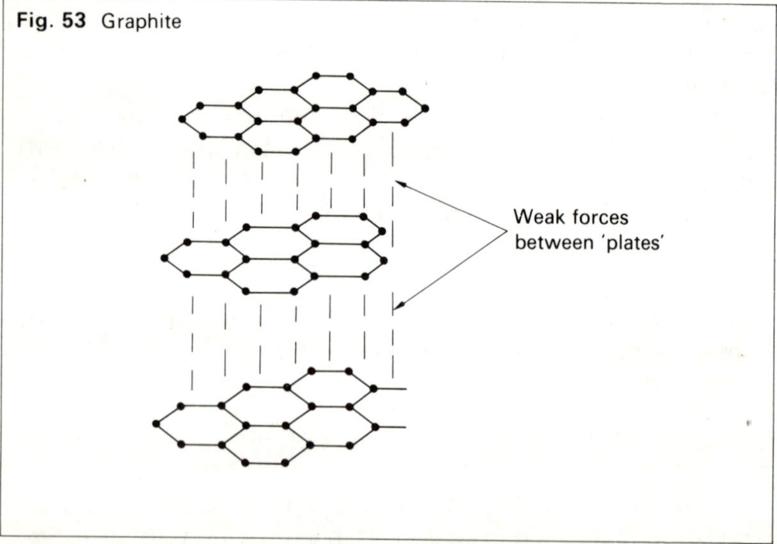

12.3 Carbon and carbon compounds as fuels

A fuel is a substance which can be used to obtain a supply of energy – usually in the form of heat.

When carbon burns in oxygen, the gas carbon dioxide is formed, and a lot of heat energy is given out. Experiments can be carried out to show the value as fuels of petrol (sooty flame when burned in air, but very hot flame as fine droplets in a spray) and petrol vapour and coal gas (explosive mixtures with air).

Fig. 54 Petrol burns with a very hot flame when mixed with air

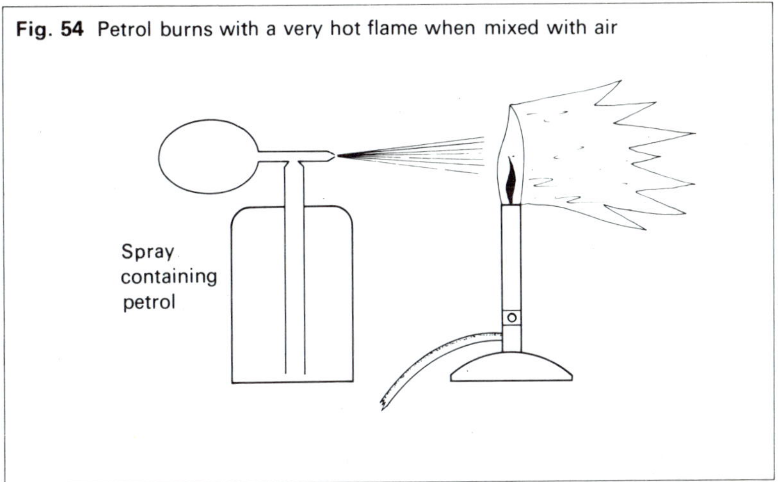

In all cases, heat energy is given off, and carbon dioxide gas is formed.

Burning carbon (in a coal or coke fire)

A number of reactions can take place.

(a) With plenty of air present:

$$C + O_2 \longrightarrow \underset{\text{carbon dioxide gas}}{CO_2}$$

(b) With a limited supply of air

$$2C + O_2 \longrightarrow \underset{\text{carbon monoxide gas.}}{2CO}$$

In a large fire, carbon monoxide is often formed, and this is seen burning with a blue flame at the surface of the fire, where more air is available.

Similar reactions take place when carbon compounds burn.

Carbon monoxide as a fuel

$$2CO + O_2 \longrightarrow 2CO_2$$

When carbon monoxide burns in air to give carbon dioxide, it releases heat energy in the process. This makes carbon monoxide a good fuel, and gases containing carbon monoxide are used as fuels in industry.

The fuel gases – producer gas and water gas

(a) When air is passed through a large mass of red-hot coke, producer gas is formed:

$$\underbrace{O_2 + 4N_2}_{\text{Air}} + 2C \longrightarrow \underbrace{2CO + 4N_2}_{\text{Producer Gas}}$$

Air $(\frac{1}{5}O_2 + \frac{4}{5}N_2)$ — Producer Gas $(\frac{1}{3}CO + \frac{2}{3}N_2)$

The nitrogen in the producer gas is, of course, of no value as a fuel.

(b) If steam is passed over red-hot coke, water gas if formed (Fig. 55)

$$C + H_2O(g) \longrightarrow \underbrace{H_2 + CO}_{\text{Water Gas}}$$

$$(50\%\ H_2 \text{ and } 50\%\ CO)$$

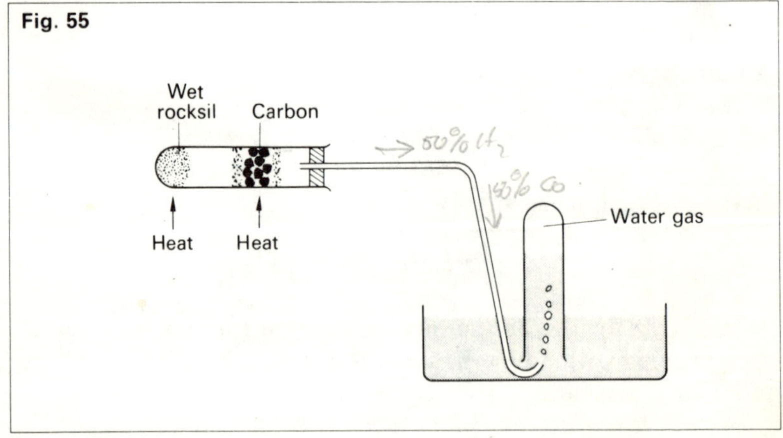

Fig. 55

Since both gases in water gas will burn, then water gas is the more efficient fuel.

Carbon monoxide is a reducing agent

Carbon monoxide will not only burn in oxygen, forming carbon dioxide, it will remove oxygen from oxygen-containing compounds, forming carbon dioxide.

Examples: $CuO + CO \longrightarrow CO_2 + Cu$

or $Fe_2O_3 + 3CO \longrightarrow 3CO_2 + 2Fe$

The second example is one of the reactions occurring in the Blast Furnace for the production of iron from its oxide ore.

Poisonous nature of carbon monoxide

Carbon monoxide reacts with the haemoglobin in the blood, preventing the haemoglobin from absorbing oxygen. Hence if no oxygen can get into the blood from the lungs, the brain cells get starved of oxygen and death quickly follows.

12.4 Carbon forms chains and rings

The carbon atom, with 4 half-filled orbitals, can form covalent bonds with other carbon atoms, forming chains. If the chain is long enough, the ends may join up to form a ring.

The remaining bonds on each carbon atom are formed with atoms of other elements such as hydrogen, oxygen or chlorine.

A compound containing only the elements hydrogen and carbon is called a hydrocarbon.

There are several series of *hydrocarbons,* called *homologous series,* the simplest one being the **alkanes.**

	Expanded structural formula	Condensed structural formula
The 1st member is methane, CH_4	H above, H—C—H, H below	CH_4

	Expanded structural formula	Condensed structural formula
The 2nd member is ethane, C_2H_6	(see below)	$CH_3.CH_3$
the 3rd member is propane, C_3H_8	(see below)	$CH_3.CH_2.CH_3$
the 4th member is butane, C_4H_{10}	(see below)	$CH_3.CH_2.CH_2.CH_3$

```
    H  H
    |  |
 H—C—C—H
    |  |
    H  H
```

```
    H  H  H
    |  |  |
 H—C—C—C—H
    |  |  |
    H  H  H
```

```
    H  H  H  H
    |  |  |  |
 H—C—C—C—C—H
    |  |  |  |
    H  H  H  H
```

If we look at butane (C_4H_{10}), we see that the 4 carbons and 10 hydrogens can join up in a different way, forming a branched chain:

```
          H
          |
        H—C—H
          |
    H     |     H
    |     |     |
  H—C—————C—————C—H
    |     |     |
    H     H     H
```

That is, $CH_3.\overset{CH_3}{\dot{C}H}.CH_3$ or, $(CH_3)_3.CH$

Molecules which have the same chemical formula, but different structural formulae, are called isomers.

The two molecules, C_4H_{10}, are isomers of butane. Try to draw the isomers of pentane (C_5H_{12}).

12.5 Hydrocarbons contain hydrogen and carbon

The products of burning a hydrocarbon such as methane are carbon dioxide (lime water turns milky) and water (b.pt. = 100°C).

Hence, the carbon in the carbon dioxide, and the hydrogen in the water must have come from the fuel being burned.

Note: To prove that all the oxygen came from the air, and that there was none present initially in the fuel before burning, some other experiment would have to be done.

Reactions of the alkanes

The alkanes are said to be *saturated* hydrocarbons (they have no double bonds), and because of this they tend to be unreactive, a hydrogen having to be pulled off before another atom can bond to the molecule.

A slow *substitution* reaction takes place, for example, with bromine:

$$\mathrm{H{-}\overset{\displaystyle H}{\underset{\displaystyle H}{C}}{-}\overset{\displaystyle H}{\underset{\displaystyle H}{C}}{-}H + Br{-}Br \longrightarrow H{-}\overset{\displaystyle H}{\underset{\displaystyle H}{C}}{-}\overset{\displaystyle H}{\underset{\displaystyle H}{C}}{-}Br + H{-}Br}$$

(white fumes)

Because the reaction is very slow; the red-brown bromine colour will persist.

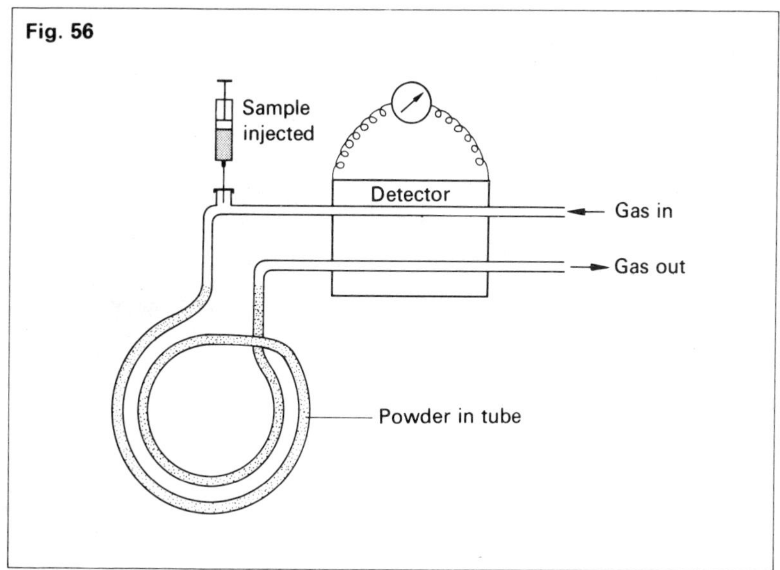

Fig. 56

12.6 Separation of hydrocarbons – gas chromatography

Since hydrocarbons are all so similar chemically, those which cannot be separated by boiling point can be separated by chromatography. (See Fig. 56.)

A gas such as nitrogen is passed through a glass tube, and the detector adjusted to give no reading on the meter.

When the sample is injected (assuming it contains a mixture of hydrocarbons) each constituent will be absorbed differently by the powder and pass through the column at a different rate, so being separated.

Each time a hydrocarbon comes off the column, it causes a change in the detector and a reading on the meter is obtained.

12.7 The Oil Industry

Oil and natural gas are usually formed from the remains of tiny marine animals. The oil and gas are trapped underground by layers of rocks.

Fig. 57

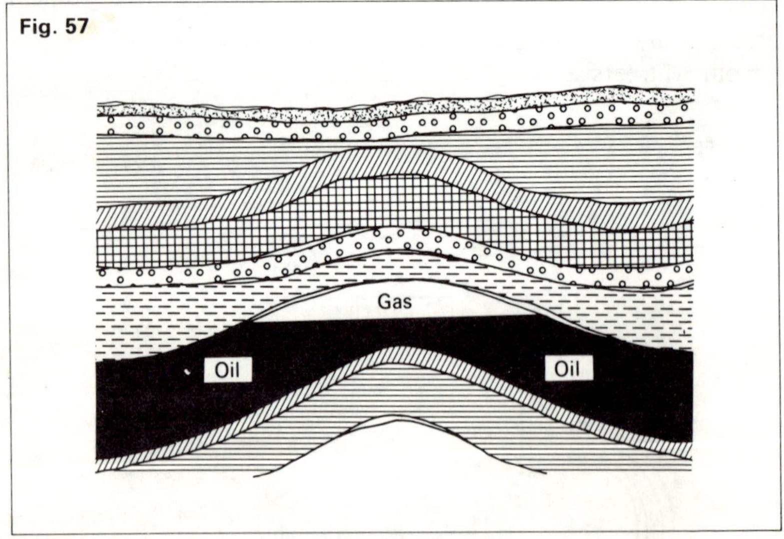

The discovery of oil and natural gas under the North Sea has been of great importance to the economy of our country, since we are very dependent on oil and natural gas for power and heating.

Natural gas (mainly methane, CH_4) is much safer than Town Gas since it does not contain poisonous carbon monoxide.

Crude oil is a dark, strongly smelling liquid. It is composed of a large number of different hydrocarbons, and the chemists' job is to separate them into useful constituents.

Fractional distillation of crude oil

This is the process by which crude oil is separated into different components, using differences in boiling point. The molecules with a small number of carbon atoms have the lowest boiling points. All the oil is heated to a high temperature, and the gases produced rise up a tower, gradually cooling down. The longer chain molecules (with high boiling points) turn into the liquid state first and are separated off. The shorter chain molecules turn into the liquid state as they cool down further up the tower.

The laboratory experiment is different from the industrial process. The temperature of the oil sample is gradually increased, and the molecules with a small number of carbon atoms boil out of the crude oil mixture at low temperatures. As the temperature is raised, molecules with more carbon atoms boil off.

Product	Carbon atoms in chain
Gases	1–4
Petrol	4–12
Kerosene (paraffin)	9–16
Diesel oil	15–25
Lubricating oil	20–70
Bitumen	more than 70

12.8 Combustion of alkanes

If we examine the ease with which different oil fractions catch fire and burn, we find that the low boiling point liquids evaporate very quickly, and catch fire without the bunsen flame touching the liquid; they have a *low ignition point.* The ease of burning steadily decreases, the higher boiling liquids not burning even when a flame is directed on to them.

Hydrocarbons as fuels

We saw earlier that hydrocarbons are used as fuels, giving off a lot of heat (and light) energy when burned in air or oxygen. The burning of ethane is an example:

$$C_2 H_6 + 7_2 + 7O_2 \longrightarrow 4CO_2 + 6H_2O + \text{energy}$$

Where does the energy come from?

1 Breaking the C – C bonds and the C – H bonds *requires* energy.

2 Making the C – O and the H – O bonds *releases* energy.

Much more energy is released in stage 2, than is required in stage 1.

The net result is that energy is released. *The reaction is exothermic.*

12.9 Putting the oil to use – thermal cracking

The amount of heavy oils which are produced (about 50% of the crude petroleum) are far greater than the demand (about 1%).

These molecules must be broken down into smaller, more useful units.

If the molecules are heated strongly, they vibrate so violently that they are broken up. This is the process of *thermal cracking.*

The cracking process may be speeded up by using a substance which gives a large, hot surface (for example, a dried clay). As a result the process is often called 'catalytic cracking' or 'cat cracking'.

Fig. 58

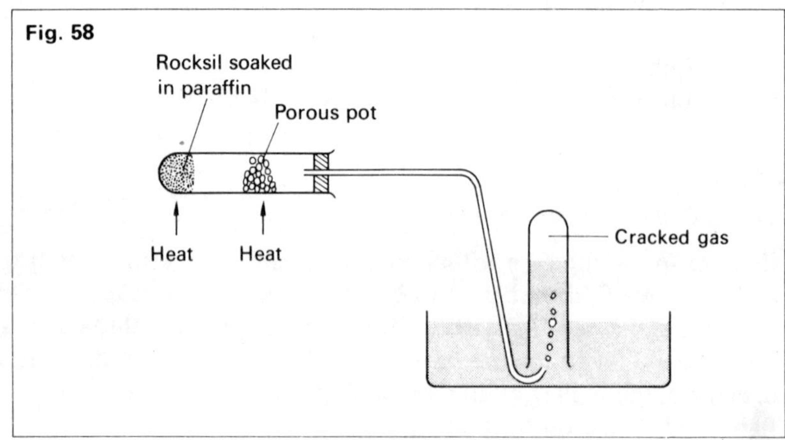

If the cracked gas was shaken with bromine water, the colour of the bromine would disappear very quickly (compare the slow reaction with alkanes).

Therefore, the cracked gases cannot all be alkanes.

What happens when molecules are cracked?

```
       H   H   H   H   H   H
       |   |   |   |   |   |
   ----C---C---C---C---C---C----
       |   |   |   |   |   |
       H   H   H   H   H   H
                 |
                 ↓
       H   H   H   H   H   H
       |   |   |   |   |   |
   ----C---C• •C---C• •C---C----
       |   |   |   |   |   |
       H   H   H   H   H   H
```

This leaves half-filled orbitals on the carbon atoms, and these molecules become stable by forming double bonds.

```
   H   H      H   H
   |   |      |   |
   C===C      C===C
   |   |      |   |
   H   H      H   H
```

Each time *one* bond is broken, two half-filled orbitals are produced and *one* double bond must be formed.

For example: $C_4H_{10} \rightarrow C_2H_6 + C_2H_4$

butane → ethane + ethene (double bond)

These double bonded compounds give a fast reaction with bromine water, and are said to be *unsaturated.*

By careful control of the cracking process, ethene (C_2H_4) can be obtained as a major product.

12.10 Reaction of bromine with saturated and unsaturated compounds

(a) *Saturated compounds*

$$\begin{array}{ccccccc} & & H & & H & & \\ & & | & & | & & \\ H & — & C & — & C & — & H \\ & & | & & | & & \\ & & H & & H & & \end{array} + Br - Br \longrightarrow \begin{array}{ccccccc} & & H & & H & & \\ & & | & & | & & \\ H & — & C & — & C & — & Br \\ & & | & & | & & \\ & & H & & H & & \end{array} + H\quad Br$$

This reaction is a *slow substitution.* The red-brown bromine colour persists.

(b) *Unsaturated compounds*

$$\begin{array}{ccc} H & & H \\ | & & | \\ C & = & C \\ | & & | \\ H & & H \end{array} + Br—Br \longrightarrow \begin{array}{ccccccc} & & H & & H & & \\ & & | & & | & & \\ Br & — & C & — & C & — & Br \\ & & | & & | & & \\ & & H & & H & & \end{array}$$

The bromine has a strong attraction for electrons, and is attracted to the electron clouds of the double bonds. One of the bonds breaks, and a *fast addition reaction* takes place. The red-brown bromine colour disappears quickly.

This test will distinguish between saturated and unsaturated compounds.

12.11 Using the ethene – addition polymerisation

Once the ethene has been formed, one of the most important processes is to rebuild it into very long chain molecules, forming polythene (polyethene).

One of the bonds in the double bond splits, and the molecules link up to form a long chain of over 1000 carbon atoms.

$$\begin{array}{ccc} H & & H \\ | & & | \\ C & = & C \\ | & & | \\ H & & H \end{array} + \begin{array}{ccc} H & & H \\ | & & | \\ C & = & C \\ | & & | \\ H & & H \end{array} \longrightarrow \begin{array}{ccccccccc} & H & & H & & H & & H & \\ & | & & | & & | & & | & \\ \text{---} & C & — & C & — & C & — & C & \text{---} \\ & | & & | & & | & & | & \\ & H & & H & & H & & H & \end{array}$$

ethene molecules — polythene

Addition polymerisation takes place when molecules of the same kind (the monomer) which have a carbon to carbon double bond link up as a result of one of the bonds in the double bond splitting to form a long chain molecule (the polymer).

Ethene is the basis of all addition polymers. By replacing the H atoms of ethene by other atoms, or groups, different polymers with different properties can be obtained. For example:

P.V.C. (Polyvinyl chloride)

$$\begin{array}{ccc} Cl & & H \\ | & & | \\ C & = & C \\ | & & | \\ H & & H \end{array} \longrightarrow \cdots \begin{array}{ccccccccccc} Cl & & H & & Cl & & H & & Cl & & H \\ | & & | & & | & & | & & | & & | \\ C & - & C & - & C & - & C & - & C & - & C \\ | & & | & & | & & | & & | & & | \\ H & & H & & H & & H & & H & & H \end{array} \cdots\cdots$$

monomer (vinyl chloride) — polymer (polyvinyl chloride)

Polypropylene (polypropene)

$$\begin{array}{ccc} CH_3 & & H \\ | & & | \\ C & = & C \\ | & & | \\ H & & H \end{array} \longrightarrow \cdots \begin{array}{ccccccccccc} CH_3 & & H & & CH_3 & & H & & CH_3 & & H \\ | & & | & & | & & | & & | & & | \\ C & - & C & - & C & - & C & - & C & - & C \\ | & & | & & | & & | & & | & & | \\ H & & H & & H & & H & & H & & H \end{array} \cdots$$

monomer — polymer

Teflon (P.T.F.E.)
(Polytetrafluoroethene)

$$\begin{array}{ccc} F & & F \\ | & & | \\ C & = & C \\ | & & | \\ F & & F \end{array} \longrightarrow \cdots\cdots \begin{array}{ccccccccccc} F & & F & & F & & F & & F & & F \\ | & & | & & | & & | & & | & & | \\ C & - & C & - & C & - & C & - & C & - & C \\ | & & | & & | & & | & & | & & | \\ F & & F & & F & & F & & F & & F \end{array} \cdots\cdots$$

monomer — polymer

Polystyrene

Note: [benzene ring] represents a C_6H_5 'ring'.

12.12 Properties of polythene and polystyrene

	Melting	Burning	Insulating properties	Uses
Polythene	Melts and bubbles round edges	Burns quietly with blue/ yellow flame – tends to drip off.	Both are electrical and heat insulators.	Electric insulation, storage sacks, piping, kitchen ware etc.
Polystyrene	Harder to melt; bubbles round edges	Yellow flame, thick black smoke.		Toys, lampshades, guitars. As foam: ceiling tiles, packing materials, etc.

Some revision questions

You should try to answer each question and then check your answer by referring to the section indicated in brackets after the question.

1 (a) What is meant by (i) polymorph, (ii) macromolecule?
(b) What are the polymorphs of carbon? How do their properties differ? (Section 12.2.)

2 (a) What is a fuel?
(b) Carbon can burn to form different gases, depending on the amount of air present. What are the gases?
(c) How are producer gas and water gas formed? What gases are present in each? Which is the better fuel, and why?
(d) Describe a reaction which demonstrates that carbon monoxide is a reducing agent. (Section 12.3.)

3 (a) What is meant by the following terms: (i) hydrocarbon; (ii) homologous series; (iii) structural formula; (iv) isomers?
(b) Draw the structural formulae for the isomers of butane. (Section 12.4.)

4 Describe the process of (i) fractional distillation, (ii) thermal cracking. (Sections 12.7 and 12.9.)

5 How could you distinguish between a saturated and an unsaturated hydrocarbon? (Section 12.10.)

6 (a) What is meant by 'addition polymerisation'?
(b) In addition polymerisation one *monomer* is involved. What is a monomer?
(c) Draw the structural formula of ethene.
(d) Draw the structural formula of polythene, showing how three ethene molecules have joined up together.
(Section 12.11.)

13

Foodstuffs and related substances

13.1 Carbohydrates

Carbohydrates are compounds containing carbon, hydrogen and oxygen with the hydrogen and oxygen in the proportions 2:1, as in water.

Carbohydrates are essentially chains of carbon atoms 'dressed' with H and OH groups:

$$
\begin{array}{ccccccc}
 & OH & & OH & & OH & & OH & \\
 & | & & | & & | & & | & \\
\cdots\cdot & C & — & C & — & C & — & C & \cdot\cdots \\
 & | & & | & & | & & | & \\
 & H & & H & & H & & H &
\end{array}
$$

Some proof of the composition is obtained from the fact that:

(*a*) combustion in oxygen produces carbon dioxide and water;

(*b*) when concentrated sulphuric acid is added to a carbohydrate (e.g. sugar) the elements of water are removed, leaving only carbon.

Comparison of starch and glucose – two carbohydrates (Fig. 59)

Fig. 59

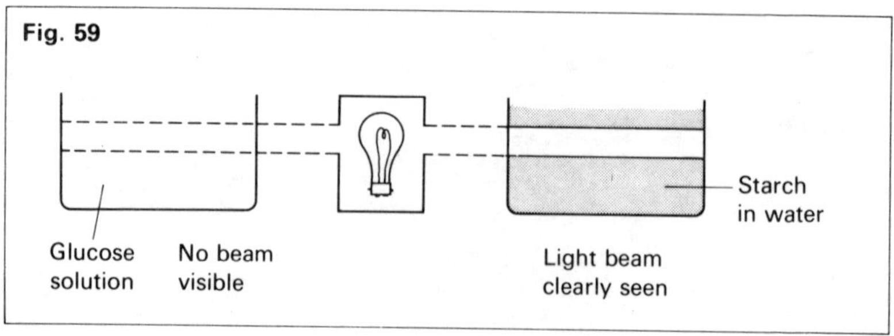

Glucose forms a true solution with water, while starch forms a colloid (its molecules must be very big – big enough to scatter light). In other words, glucose appears to be composed of small molecules, while starch has very large molecules.

Classes of carbohydrate

1 *Monosaccharides (simple sugars)*. For example: glucose and fructose. They have the formula $C_6H_{12}O_6$ (a chain of six carbon atoms).
2 *Disaccharides* (complex sugars). For example: maltose and sucrose. They have the formula $C_{12}H_{22}O_{11}$ (a chain of 12 carbon atoms).
3 *Polysaccharides* (non-sugars). Very long chain molecules of formula $(C_6H_{10}O_5)_n$ where n = approximately 300 for starch and approximately 3000 for cellulose.

13.2 Tests for saccharides

1 *All sugars except sucrose*

React with Fehling's solution, or Benedict's reagent to give an orange-red precipitate. These sugars are called *reducing sugars.*

2 *Starch*

Reacts with iodine, giving a blue-black colour.

If the above tests are carried out on plants and foodstuffs, the presence of starch and reducing sugars is shown in many of them.

13.3 Connection between starch and glucose (hydrolysis of starch)

In the presence of acid, starch molecules can be broken down, by combining with water, to a reducing sugar. (The reducing sugar formed is, in fact, glucose.) Hydrochloric acid is used to help the reaction.

$$\underset{\text{starch}}{(C_6H_{10}O_5)_n} + n\,H_2O \rightarrow \underset{\text{glucose}}{n\,C_6H_{12}O_6}$$

The starch has been *hydrolysed.*

Hydrolysis is the process by which molecules are broken down by addition of water.

This suggests that starch is made up by the polymerisation of a large number of glucose molecules (water being lost in the process):

A polymerisation where water is lost in the process, is called a condensation polymerisation.

$$\underset{\text{glucose}}{n\ C_6H_{12}O_6} \underset{\text{hydrolysis}}{\overset{\text{condensation}}{\rightleftharpoons}} \underset{\text{starch}}{(C_6H_{10}O_5)_n} + n\ H_2O$$

13.4 Hydrolysis of starch in our bodies

Starch is broken down to the reducing sugar, maltose, by saliva. This can be shown by experiment, using a sample of your own saliva.

The breakdown is carried out by a catalyst called an *enzyme.*

An enzyme is a big molecule with a definite shape which allows it to carry out a particular reaction (to build up, or break down a molecule). An enzyme is a specialised catalyst which can only do one job.

In the breakdown of starch by saliva, the enzyme *amylase* (sometimes called *ptyalin*) in the saliva assists the breakdown process.

13.5 Carbohydrates as energy sources

When carbohydrates are burned in oxygen they produce a great amount of energy. A mixture of air and a fine powder such as flour, is explosive. Carbohydrates are used as energy sources in, for example, the burning of paper, or wood in a fire.

Within the body, the same amount of energy is released but, of course, very much more slowly at the normal body temperature of 37°C. We eat carbohydrates, which we store and 'burn' up at the required time to give us energy, the process being called *respiration.* Plants must manufacture their own carbohydrates, and do this by the process of *photosynthesis,* energy being taken in during the process.

13.6 Photosynthesis

Carbohydrates are made in plants from carbon dioxide (in the air) and water (from the soil). Light energy is supplied from the sun, and chlorophyll (the green colouring matter) is required to help the process. Since light energy is required, the process is called photosynthesis.

$$6\,CO_2 + 6\,H_2O + \begin{matrix}\text{light}\\\text{energy}\end{matrix} \xrightarrow[\text{(in presence of chlorophyll)}]{\text{photosynthesis}} \begin{matrix}C_6H_{12}O_6 + 6\,O_2\\\text{monosaccharide}\\\text{(e.g. glucose)}\end{matrix}$$

Oxygen gas is released into the air in this process. The glucose formed is then polymerised to starch and cellulose, and the starch is stored until required.

Photosynthesis is the process by which plants manufacture carbohydrates from carbon dioxide and water, using light energy in the presence of chlorophyll. Oxygen is released in the process.

13.7 Respiration

All living cells, whether plants or animals, require energy, and this energy is obtained by the 'combustion' of carbohydrates. The starch is broken down (*hydrolysed*) to glucose, and the glucose (using oxygen) is broken down into carbon dioxide and water, releasing energy.

This process is *respiration*, and is the reverse of photosynthesis.

$$C_6H_{12}O_6 + 6\,O_2 \underset{\text{photosynthesis}}{\overset{\text{respiration}}{\rightleftharpoons}} 6\,CO_2 + 6\,H_2O + \text{Energy}$$

Respiration is the process by which plants and animals obtain a supply of energy by breaking down carbohydrates (using oxygen) into carbon dioxide and water, the energy being released in the process.

Respiration in Animals

This is summarized in the chart below.

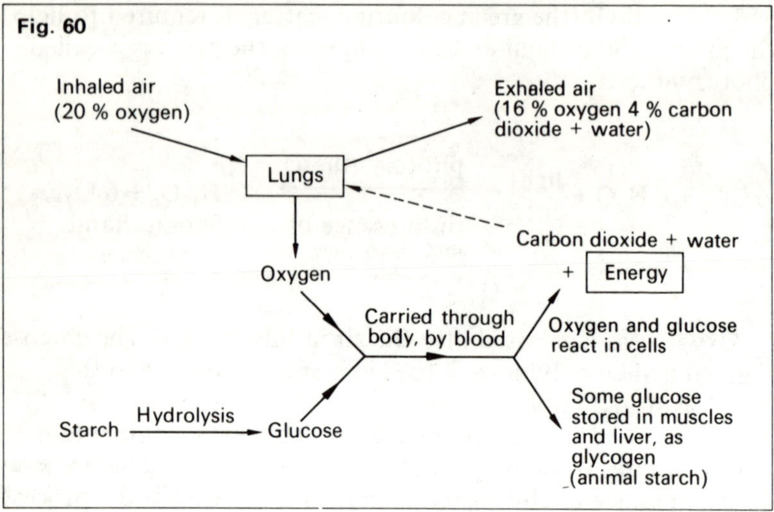

13.8 The Carbon Cycle

This is best summarized in the chart below.

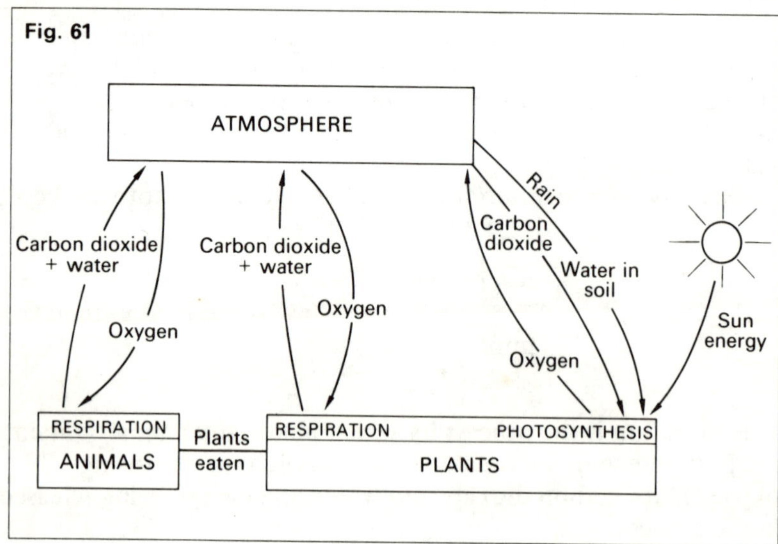

13.9 Disaccharides

Sucrose is the only non-reducing sugar (the only sugar which doesn't react with Fehling's or Benedict's solutions).

Hydrolysis of sucrose

When sucrose is hydrolysed using acid, reducing sugars are obtained.

$$\underset{\text{Sucrose}}{C_{12}H_{22}O_{11}} + H_2O \longrightarrow \underset{\text{Reducing Sugars}}{2C_6H_{12}O_6}$$

The presence of the reducing sugars is shown by Fehling's test. (In this test, the deep blue colour of Fehling's solution changes to green, and then a reddish-brown precipitate of copper(I) oxide forms.)

Identifying the products of hydrolysis of sucrose – chromatography

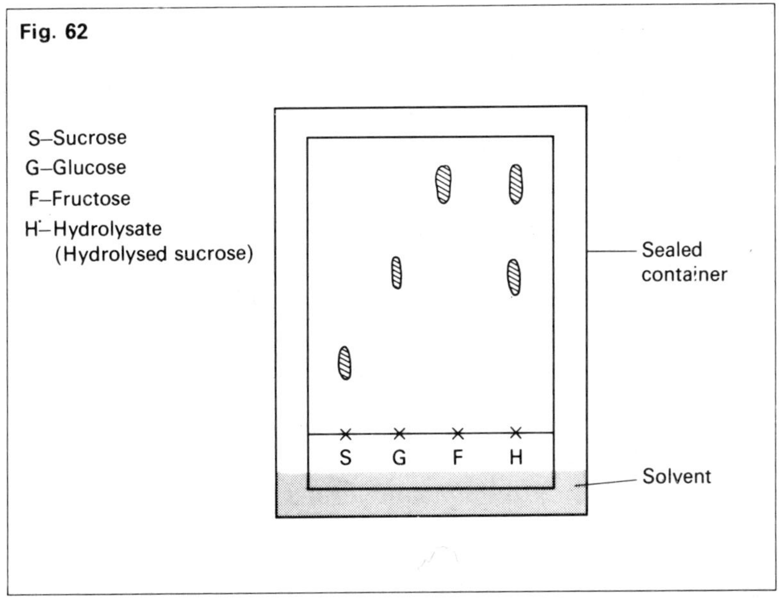

Known sugars and the hydrolysate (hydrolysed sucrose) are spotted on the paper in a line at the points marked 'X' (Fig. 62). As the liquid soaks up the paper, it washes the carbohydrates along at different rates.

When the liquid has moved up the paper, the paper is dried and the position of the spots shown by using a spray which reacts with the sugars.

In this case the hydrolysed sucrose is shown to contain glucose and fructose, since it separates into two spots which moved the same distance as the glucose and fructose samples.

This shows that: $$\text{Glucose} + \text{Fructose} \underset{\text{Hydrolysis}}{\overset{\text{Condensation}}{\rightleftharpoons}} \text{Sucrose.}$$

Hydrolysis of Maltose shows the following results:

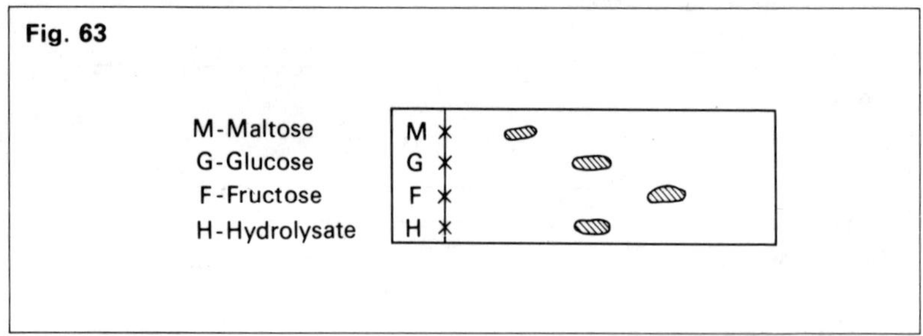

Fig. 63

13.10 Alcohols

There exists a whole family of alcohols, but the one most often referred to is *ethanol*, which is of great importance to the brewing industry.

We have seen how starch can be broken down to glucose by enzymes. But these sugars can be broken down further by another series of enzymes into simpler molecules, one of these molecules being ethanol (C_2H_5OH).

This reaction is carried out when there is insufficient air for complete respiration. The process is a special kind of respiration.

13.11 The Brewing industry

The process can be summarized as taking place in the following steps.

1 Barley shoots are ground up to obtain the enzyme, amylase (sometimes called diastase).

2 This is added to starchy material, and

$$\text{starch} \xrightarrow{\text{amylase}} \text{maltose}$$

3 Yeast is now added and the yeast supplies enzymes, called maltase, and zymase.

These catalyse the following reactions:

(a) $\underset{\text{maltose}}{C_{12}H_{22}O_{11}} + H_2O \xrightarrow{\text{maltase}} \underset{\text{glucose}}{2\,C_6H_{12}O_6}$

(b) $\underset{\text{glucose}}{C_6H_{12}O_6} \xrightarrow{\text{zymase}} \underset{\text{ethanol}}{2C_2H_5OH} + 2CO_2$

4 The result is a dilute ethanol solution, which can be fractionally distilled to yield a solution containing about 95% ethanol.

Note: You have now come across two other names for *amylase.*

1 In saliva, it used to be called *ptyalin.*

2 In plants, it used to be called *diastase.*

The names ptyalin and diastase are still used occasionally.

Fermentation of glucose into ethanol

The reaction is similar to the final stages of the brewing industry reactions above, and can readily be carried out in the laboratory.

13.12 Polar nature of ethanol

```
   H  H                                    H  H
   |  |                                    |  |     δ− δ+
H—C—C—H     One 'H' replaced by 'OH'    H—C—C—O—H
   |  |                                    |  |
   H  H                                    H  H
```

ethane (non polar) ethanol (polar molecule)

The oxygen in the ethanol, having a strong attraction for electrons, makes the molecule polar and therefore gives ethanol different properties from ethane.

13.13 Properties of ethanol

(a) Ethanol is completely soluble in water.
(b) Its pH = 7, and it is a non-conductor of electricity.
(c) It burns, giving off carbon dioxide and water.

$$C_2H_5OH + 3O_2 \rightarrow 2CO_2 + 3H_2O$$

(Note: This experiment would only prove C and H to be present in ethanol. Since oxygen was used in burning, all the oxygen in the products could have come from the air.)
(d) Ethanol may be oxidised to ethanoic acid (acetic acid). If wine is not stored properly it goes sour, due to the ethanol being oxidised to ethanoic acid. Industry uses large amounts of ethanoic acid, and it can be made from ethanol chemically. Ethanol vapour reacts with hot copper (II) oxide, copper being formed and ethanoic acid vapour being produced (shown by pH indicator solution turning red).

Fig. 64

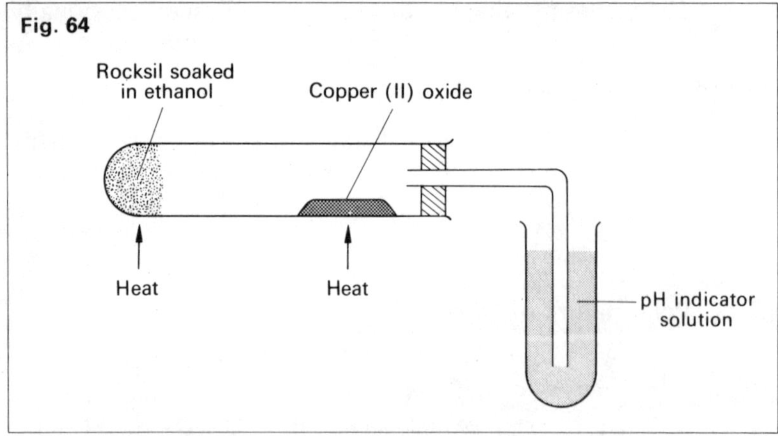

$$2CuO + CH_3CH_2OH \rightarrow 2Cu + CH_3COOH + H_2O$$

ethanoic acid

13.14 Ethanoic acid CH_3COOH

Structure

```
     H      O
     |     //
H—C—C
     |     \
     H      O—H
```

Properties

(a) Soluble in water, having a pH less than 7. Therefore it is acidic. The pH of dilute solutions is usually found around 3; therefore, it is a *weak acid.*
(b) Conductivity is very low compared with acids like hydrochloric acid. This confirms that it is a weak acid.
(c) It shows normal acid properties; for example, it reacts with magnesium to produce hydrogen.

Ethanoic acid – a weak acid

When dissolved in water, only a few of its molecules dissociate into ions. Therefore, at any one instant, only a small proportion of acid is present as ions.

$CH_3COOH + H_2O \rightleftharpoons$	$CH_3COO^-(aq) + H^+(aq)$
most of the solution is in this form	only a few ions free at any instant

13.15 A series of acids

There is a whole series of acids. Each one has a similar formula and name based on the alkane with the corresponding number of carbon atoms; the name ends in '-oic acid'.

For example:	$HCOOH$	methanoic acid
	$CH_3.CH_2.COOH$	propanoic acid

13.16 Reaction of acids with alcohols – esters

Esters are formed as a result of a condensation reaction (removal of water) between an acid and an alcohol. Usually concentrated sulphuric acid is used to assist the removal of water, and the mixture is heated on a water bath for some time.

Two examples are given below:

$$CH_3.COOH + HOCH_2.CH_3$$

ethanoic acid — ethanol

condensation ↓ (H_2O removed)

$$CH_3.COOCH_3.CH_3 + H_2O$$

ethyl ethonoate (an ester)

$$CH_3.COOH + HOCH_3$$

ethanoic acid — methanol

condensation ↓ (H_2O removed)

$$CH_3.COO.CH_3 + H_2O$$

methyl ethanoate (an ester)

Naming of Esters

The name of an ester depends on the acid and alcohol from which it is prepared.

For example:–

$CH_3 \cdot CH_2 \cdot COO$	$CH_2 \cdot CH_3$
Name based on 'parent' acid – propanoic acid.	Name based on 'parent' alcohol – ethanol.
Therefore named *propanoate*	Named *ethyl*

Therefore name of ester: *ethyl propanoate.*

Hydrolysis of esters is the reverse of *condensation,* the ester being broken up into the acid and alcohol.

Uses of esters

(1) Nail varnish, and nail varnish remover.
(2) Car paint thinners.
(3) Artificial flavourings.
(4) Perfumes and cosmetics (esters have a 'fruity' smell).

13.17 Fats and oils

All fats and oils are of vegetable or animal origin. They are *esters* of the polyalcohol (an alcohol with a number of OH groups) called glycerol.

The acids forming the esters are mainly the liquid oleic acid or the solid stearic acid.

$$C_{17}H_{35}\cdot COOH \;+\; -\overset{|}{\underset{|}{C}}-OH \xrightarrow{\text{condensation}} C_{17}H_{35}\cdot COO\overset{|}{\underset{|}{C}}- \;+\; H_2O$$

Stearic acid — part of glycerol molecule — a fat

$$C_{17}H_{33}\cdot COOH \;+\; -\overset{|}{\underset{|}{C}}-OH \xrightarrow{\text{condensation}} C_{17}H_{33}\cdot COO\overset{|}{\underset{|}{C}}- \;+\; H_2O$$

oleic acid — an oil

Fats and oils as energy sources

Fats are important energy sources, although it is doubtful if they are essential for the release of energy in animals.

Fats give out more energy than the same weight of carbohydrates, and for this reason people in cold climates eat much more fat than others in warmer countries.

When fats are digested, the first breakdown results in the formation of glycerol and the acid. These are transported through the body, where they either break further, releasing energy, or they recombine to form fat.

Oils are unsaturated

Oils in animals and plants (for example, whale oil, coconut oil, olive oil) are liquid at ordinary temperatures. They are formed from the unsaturated oleic acid, and will decolourise bromine water. (See test for unsaturated compounds, section 12.10.)

13.18 Margarine manufacture

The oils are 'hardened' by adding hydrogen across the double bonds, using a catalyst. The 'hardened' material is then mixed with the correct amount of unsaturated oil until the right consistency is achieved.

13.19 Soap is manufactured from fats and oils

When fats or oils are hydrolysed by boiling with a solution of an alkali (for example, sodium hydroxide), glycerol and an acid, such such as stearic acid, are produced. The acid is immediately neutralized by the alkali to form the salt, sodium stearate (a soap).

$$\text{Fat} + \text{sodium hydroxide} \xrightarrow[\text{neutralization}]{\text{hydrolysis and}} \text{sodium stearate (soap)} + \text{glycerol} + \text{water}$$

The soap and glycerol are separated by 'salting' – adding sodium chloride solution.

The soap molecule consists of two parts: the long chain $C_{17}H_{35}$, which is covalent, and the remainder, $COO^- Na^+$, which is ionic.

Fig. 65

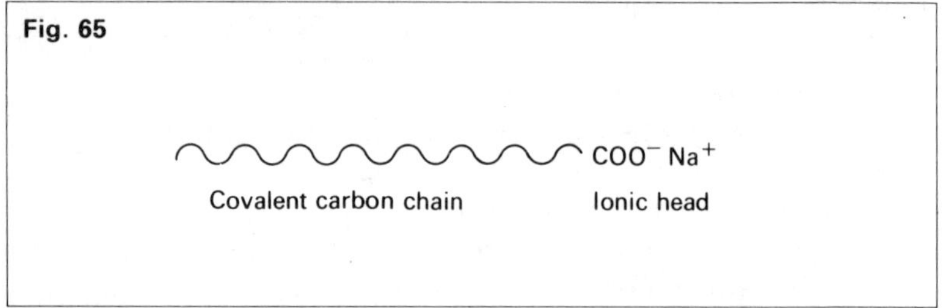

13.20 The cleansing (detergent) action of soap

A soap molecule has a covalent tail (insoluble in water, soluble in oil) and an ionic head (soluble in water, insoluble in oil). (Fig. 65.)

Oil and water don't mix. If soap molecules are added they line up as in Fig. 66(a). When the mixture is shaken soap molecules form round each oil droplet as in Fig. 66(b).

The positive sodium ions move into the water, leaving a series of negative charges as in Fig. 66(c). This prevents the droplets coming together, and so forms an *emulsion.*

Fig. 66

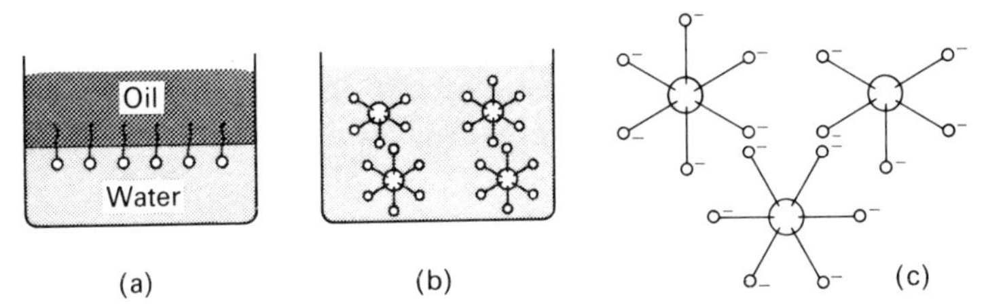

Washing the dishes

In removing grease from a plate, soap acts in exactly the same way, forming an emulsion of the grease in water.

13.21 The action of soap on hard water

In some areas of Britain, the water in the reservoirs contains calcium ions (Ca^{2+}) or magnesium ions (Mg^{2+}). This water is called *hard water,* since it forms a scum with soap instead of forming a lather. The scum is insoluble calcium stearate.

Sodium stearate + calcium ions → Calcium stearate + sodium ions
(soap) (insoluble scum)

Solving the problem of hard water

1 The calcium or magnesium ions can be removed chemically. This can be expensive.
2 **Soapless detergents** may be used. Chemists have produced a molecule which acts like a soap but does not react with calcium or magnesium ions. The petroleum industry can supply an abundance of long chain hydrocarbons, and a long hydrocarbon chain, ending in a ring of six carbon atoms is chosen. When this is treated with oleum (super concentrated sulphuric acid) it takes on a suitable head group, which, when neutralized, gives the detergent (Fig. 67).

Fig. 67

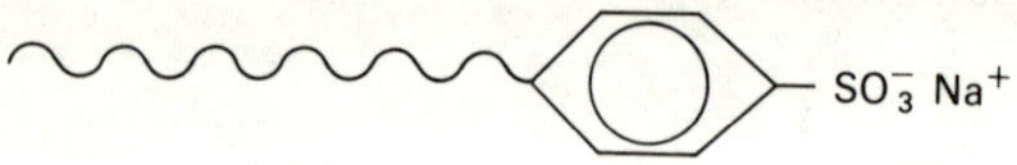

$$\sim\sim\sim\sim\sim C_6H_4 - SO_3^- \, Na^+$$

The calcium or magnesium salts of this detergent are soluble in water, and therefore the detergent gives a lather with hard water.

What is a detergent?

Soaps or soapless detergents are *all* classified as detergents, because they have the same cleansing action. (See 13.20.)

13.22 Proteins

Proteins are another group of long chain molecules. They are the 'building material' of nature.

Common protein foods are meat, eggs, fish, etc. Other protein materials are skin, hair, fingernails, wool, gelatin.

Composition of proteins

When a protein is heated with solid sodium hydroxide, a gas comes off which smells like ammonia, and turns pH paper blue (that is, the gas is alkaline).

Thus, a protein is related in some way to ammonia.

13.23 Amino acids, the building bricks

ammonia	amine	amino acid
H_2N-H	$H_2N-CH_2-CH_3$	$H_2N-CH(CH_3)-COOH$
a base (alkaline in water)	a base	basic group (H_2N-), acid group ($-COOH$)

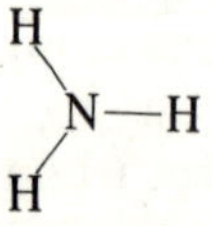

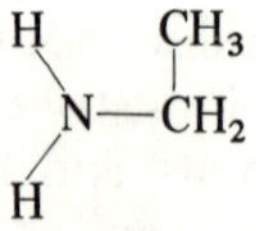

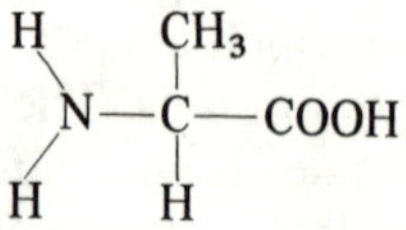

There are about 24 amino acids, which make up all natural proteins by different combinations. Amino acids only differ in the length of the carbon chain, and may be represented:–

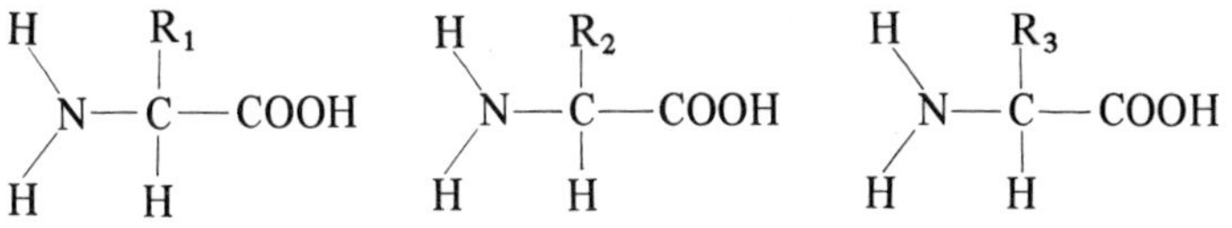

$$\begin{array}{ccc} \overset{H}{\underset{H}{}} \!\!\!\!\!\! \begin{array}{c} \diagdown \\ N \\ \diagup \end{array} \!\!-\!\! \underset{H}{\overset{R_1}{\underset{|}{\overset{|}{C}}}} \!\!-\!\! COOH & \overset{H}{\underset{H}{}} \!\!\!\!\!\! \begin{array}{c} \diagdown \\ N \\ \diagup \end{array} \!\!-\!\! \underset{H}{\overset{R_2}{\underset{|}{\overset{|}{C}}}} \!\!-\!\! COOH & \overset{H}{\underset{H}{}} \!\!\!\!\!\! \begin{array}{c} \diagdown \\ N \\ \diagup \end{array} \!\!-\!\! \underset{H}{\overset{R_3}{\underset{|}{\overset{|}{C}}}} \!\!-\!\! COOH \end{array}$$

(R_1, R_2 and R_3 are carbon chains of different lengths.)

Amino acids condense to form proteins

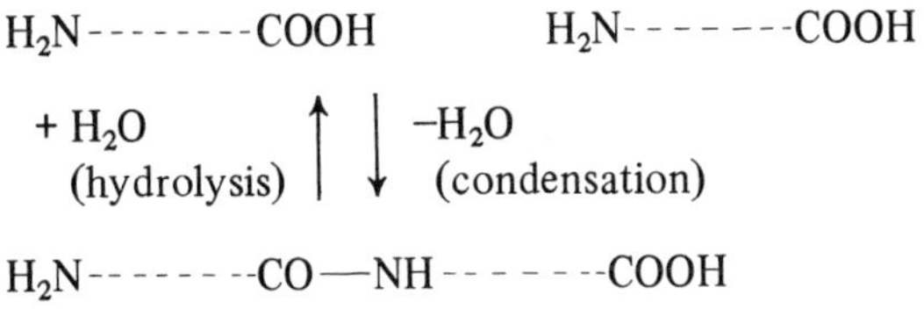

H_2N--------COOH $\qquad$ H_2N-------COOH

$+ H_2O$ (hydrolysis) ↑↓ $-H_2O$ (condensation)

H_2N-------CO—NH-------COOH

The CO–NH link formed in the condensation is called the Peptide Link.

A protein molecule is formed when a large number of the amino acid molecules join up together.

13.24 Hydrolysis of proteins and detection of the amino acids formed

Proteins can be hydrolysed (broken down) in a test tube by heating with hydrochloric acid.

The amino acids can be separated, and detected by paper chromatography.

13.25 Digestion of proteins

The enzyme *pepsin* digests (hydrolyses) the protein into its amino acids. These are absorbed in the blood and transported to part of the body which is growing, or in need of repair. There, the amino acids are reassembled (condensed) into a different order, to make the proteins required for body tissue.

Some revision questions

You should try to answer each question, and then check your answer by referring to the section indicated in brackets after the question.

1 Name the three classes of carbohydrate and give the general formula for each class. (Section 13.1.)

2 (a) What is the test for a reducing sugar? Which sugar will *not* give a positive test?
(b) What is the test for starch?
(c) How could you show that starch was built up from reducing sugars?
(d) What is meant by the word *hydrolysis*?
(Sections 13.1 to 13.4.)

3 Describe the processes of *photosynthesis* and *respiration.* (Section 13.6 and 13.7.)

4 How could you distinguish between glucose and fructose? (Section 13.9.)

5 Describe how you could prepare a sample of ethanol starting from glucose. (Section 13.11.)

6 Give an equation, and describe how you could oxidise ethanol to ethanoic acid. (Section 13.13.)

7 What is an ester, and how could you make one? Give an example. (Section 13.16.)

8 (a) Describe how fat can be converted into soap.
(b) How can soap be used to remove grease from clothing?
(c) What is a soapless detergent? Why is it more useful in hard water areas?
(Sections 13.19 to 13.21.)

9 (a) What is the 'building brick' of a protein?
(b) Show how any three of these 'building bricks' combine together to form a protein.
(Section 13.23.)

14

Macromolecules

A macromolecule is a very big molecule in which a very large number of the atoms are joined by covalent bonds.

Polymerisation is the process where a number of simple molecules called monomers join up together, usually forming a long chain. A macromolecule is formed.

There are two ways by which small molecules can be linked with each other to give larger ones.

(1) Addition polymerisation.

Here only one type of monomer is used. The monomer is based on ethene ($CH_2 = CH_2$), and they join together by breaking one of the carbon–carbon bonds, and linking head to tail.

This was covered in Section 12.11.

$$\text{e.g.}\quad \begin{array}{ccc} H & & H \\ | & & | \\ C & = & C \\ | & & | \\ H & & H \end{array} \; + \; \begin{array}{ccc} H & & H \\ | & & | \\ C & = & C \\ | & & | \\ H & & H \end{array} \; + \; \begin{array}{ccc} H & & H \\ | & & | \\ C & = & C \\ | & & | \\ H & & H \end{array} \rightarrow$$

$$\begin{array}{ccccccccccccc} & H & & H & & H & & H & & H & & H & \\ & | & & | & & | & & | & & | & & | & \\ \ldots\ldots & C & - & C & - & C & - & C & - & C & - & C & \ldots\ldots \\ & | & & | & & | & & | & & | & & | & \\ & H & & H & & H & & H & & H & & H & \end{array}$$

Rubber – a natural addition polymer

The chains in rubber are similar to polythene, but double bonds are still present. These double bonds can be attacked by atmospheric oxygen and the rubber becomes perished.

Vulcanising – Sulphur atoms link between chains of rubber molecules, cutting down the double bonds and at the same time making it less flexible.

(2) Condensation polymerisation

The monomers join together by elimination of water.
Natural polymers are usually formed by condensation reactions – examples are starch and cellulose (condensation of glucose) and proteins (condensation of amino acids).

In *some* natural condensation polymerisations, one monomer only is involved, for example glucose polymerising to starch. In others (e.g. protein formation) a number of monomers are involved.

Synthetic condensation polymerisation

Two different monomers are involved, and water is eliminated between the molecules:

$$HO—A—OH \quad H—B—H \quad HO—A—OH \quad H—B—H$$
$$\downarrow -H_2O$$
$$---A-B-A-B---$$

14.1 Synthetic condensation polymers

These can be derived from two main sources:
(a) converting existing polymers such as cellulose (cotton wool) into more suitable structures such as rayon;
(b) by building up new polymers from a variety of monomers, (for example terylene, urea-formaldehyde).

Nylon is made from two monomers:
(1) a string of carbon atoms (X) with an acid group at each end

$$HOOC–X–COOH$$

(2) a string of carbon atoms (Y) with an amine (— NH_2) group at each end

$$H_2N–Y–NH_2$$

The NH_2 group (as in amino acids) is related to ammonia, and has basic properties.

Condensation occurs between the acidic and basic groups.

$$HOOC \;\; X \;\; COOH \quad H_2N—Y—NH_2 \quad HOOC—X—COOH \quad H_2N—Y—NH_2$$
$$\downarrow -H_2O$$
$$---X—CO\cdot NH—Y—NH\cdot CO—X—CO\cdot NH—Y---$$

In the laboratory, it is easy to make nylon if we have an acid chloride group (–COCl) at each end of the molecule 'X', instead of an acid group.

An experiment in which nylon is formed is shown in Fig. 68.

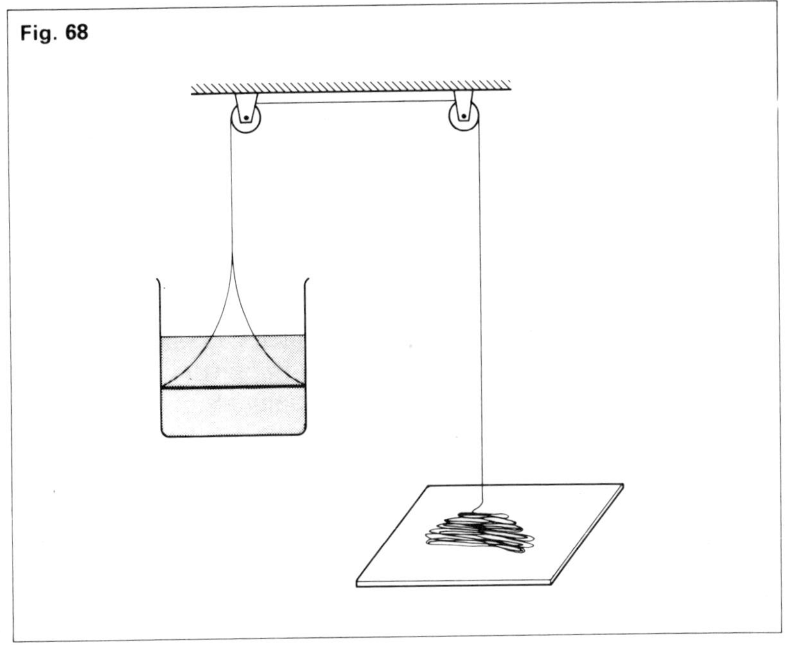

Fig. 68

The top layer is a diamine ($N_2N-Y-NH_2$) dissolved in sodium hydroxide solution. The bottom layer is a diacid chloride (ClOC–X–COCl) dissolved in tetrachloromethane.

Various kinds of Nylon can be obtained, depending on the length of carbon chains.

Animal fibres, such as hair, wool and silk are proteins. The artificial fibre, nylon, has also the properties of protein – it gives off ammonia with sodium hydroxide pellets.

14.2 Thermoplastics and thermosetting plastics

Nylon can be melted and cooled back to its original form. It is said to be a *thermoplastic.* Thermoplastics can be melted and re-shaped over and over again. Examples are nylon, terylene, perspex, polythene, P.V.C., polystyrene, P.T.F.E.

They are usually composed of long chain molecules, which when cold, are tangled up together.

Fig. 69

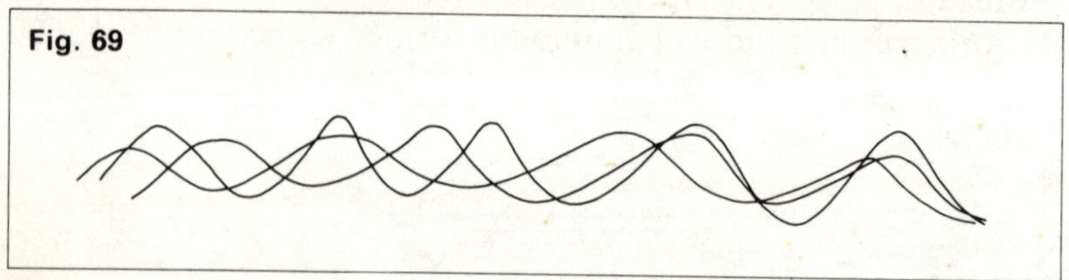

Heating untangles the chains, due to the increased vibrations of the molecules, and the plastic becomes more pliable and melts.

Urea-formaldehyde can be shaped and baked to form a hard solid which does not melt easily. It is usually called Bakelite, and is said to be a *thermosetting* plastic.

Thermosetting plastics are plastics which harden on heating and do not melt on reheating. Other thermosetting plastics are epoxy resins (such as Araldite) and melamine-formaldehyde (Formica). Heating a thermosetting plastic causes the chains to cross-link, making the plastic rigid, and difficult to melt (Fig. 70).

Fig. 70

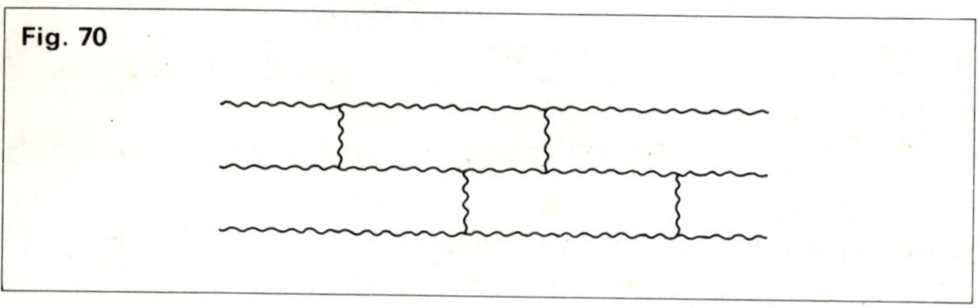

14.3 Identification of plastics by burning

Most plastics burn in different manners and it is possible to identify some of them by their burning properties.

Substance	Burning property
Nylon	Blue flame with yellow tip and no smoke. Smells of burning wool. The material drips and the flame tends to go out.
Cellulose Acetate	Yellow flame, some light smoke. There is a smell of vinegar and often spurting.
Polythene	Burns quietly, blue/yellow flame — tends to drip.
Polystyrene	Yellow flame, thick black smoke — drips.
PTFE	Does not burn. Feels greasy when cold.

14.4 Silicones

Silicon is an element in group 4 of the Periodic Table and, like carbon, can form chains.

Silicon atoms link alternately with oxygen atoms, and the other silicon bonds are attached to methyl groups ($-CH_3$).

```
        CH3       CH3       CH2       CH3
         |         |         |         |
---O—Si—O—Si—O—Si—O—Si—O---
         |         |         |         |
        CH3       CH3       CH3       CH3
```

Varying the length of the Si–O– chain, and introducing cross linking between chains, gives a great variety of silicones, from oily liquids through waxes and rubbery plastics to hard solids.

Groups other than the CH_3- group are sometimes introduced to give even more variation in properties.

Properties and uses of silicones

1 Their water repellant and anti-stick properties are used in polishes and non-stick linings.

2 They are good insulators of heat and electricity, and can withstand high temperatures. Silicone rubbers can withstand heat and petrol, where ordinary rubbers and plastics do not. Silicones can also be used in electrical insulation of motors which have to withstand heat and water.

To prevent short circuits, the electronics industry insulates micro circuits by coating them in silicones.

Some revision questions

You should try to answer each question, and then check your answer by referring to the section indicated in brackets after each question.

1 What is meant by *addition polymerisation*? Give an example. (See introductory section.)

2 What is meant by *condensation polymerisation*? Give an example. (See introductory section, and 14.1.)

3 What is (i) a thermoplastic, (ii) a thermoset? (Section 14.2).

4 Draw part of a silicone chain. Why are silicones valuable? (Section 14.4).

Appendix: Data tables

Periodic Table

Symbol	Atomic Number Atomic Weight
Electron arrangement Density (kg m^{-3})* Name	

Col. 1	Col. 2
H 1 1·01 1 0·0899 Hydrogen	
Li 3 6·94 2, 1 0·53 × 10^3 Lithium	**Be** 4 9·01 2, 2 1·85 × 10^3 Beryllium
Na 11 23·0 2, 8, 1 0·97 × 10^3 Sodium	**Mg** 12 24·3 2, 8, 2 1·74 × 10^3 Magnesium

TRANSITION ELEMENTS

Col. 1	Col. 2							
K 19 39·1 2, 8, 8, 1 0·86 × 10^3 Potassium	**Ca** 20 40·1 2, 8, 8, 2 1·54 × 10^3 Calcium	**Sc** 21 45·0 2, 8, 9, 2 2·99 × 10^3 Scandium	**Ti** 22 47·9 2, 8, 10, 2 4·50 × 10^3 Titanium	**V** 23 50·9 2, 8, 11, 2 5·96 × 10^3 Vanadium	**Cr** 24 52·0 2, 8, 13, 1 7·20 × 10^3 Chromium	**Mn** 25 54·9 2, 8, 13, 2 7·20 × 10^3 Manganese	**Fe** 26 55·8 2, 8, 14, 2 7·86 × 10^3 Iron	**Co** 27 58·9 2, 8, 15, 2 8·90 × 10^3 Cobalt
Rb 37 85·5 2, 8, 18, 8, 1 1·53 × 10^3 Rubidium	**Sr** 38 87·6 2, 8, 18, 8, 2 2·60 × 10^3 Strontium	**Y** 39 88·9 2, 8, 18, 9, 2 4·47 × 10^3 Yttrium	**Zr** 40 91·2 2, 8, 18, 10, 2 7·09 × 10^3 Zirconium	**Nb** 41 92·9 2, 8, 18, 11, 2 8·57 × 10^3 Niobium	**Mo** 42 95·9 2, 8, 18, 13, 1 10·2 × 10^3 Molybdenum	**Tc** 43 98·9 2, 8, 18, 14, 1 11·5 × 10^3 Technetium	**Ru** 44 101 2, 8, 18, 15, 1 12·3 × 10^3 Ruthenium	**Rh** 45 103 2, 8, 18, 16, 1 12·4 × 10^3 Rhodium
Cs 55 133 2, 8, 18, 18, 8, 1 1·88 × 10^3 Caesium	**Ba** 56 137 2, 8, 18, 18, 8, 2 3·51 × 10^3 Barium	**La** 57 139 2, 8, 18, 18, 9, 2 6·19 × 10^3 Lanthanum	**Hf** 72 178 2, 8, 18, 32, 10, 2 13·3 × 10^3 Hafnium	**Ta** 73 181 2, 8, 18, 32, 11, 2 16·6 × 10^3 Tantalum	**W** 74 184 2, 8, 18, 32, 12, 2 19·4 × 10^3 Tungsten	**Re** 75 186 2, 8, 18, 32, 13, 2 20·5 × 10^3 Rhenium	**Os** 76 190 2, 8, 18, 32, 14, 2 22·5 × 10^3 Osmium	**Ir** 77 192 2, 8, 18, 32, 17, 0 22·4 × 10^3 Iridium
Fr 87 2, 8, 18, 32, 18, 8, 1 Francium	**Ra** 88 226 2, 8, 18, 32, 18, 8, 2 5·00 × 10^3 Radium	**Ac** 89 2, 8, 18, 32, 18, 9, 2 Actinium						

LANTHANIDES:	**La** 57 139 2, 8, 18, 18, 9, 2 Lanthanum	**Ce** 58 140 2, 8, 18, 20, 8, 2 Cerium	**Pr** 59 140 2, 8, 18, 21, 8, 2 Praseodymium	**Nd** 60 144 2, 8, 18, 22, 8, 2 Neodymium	**Pm** 61 2, 8, 18, 23, 8, 2 Promethium	**Sm** 62 150 2, 8, 18, 24, 8, 2 Samarium
ACTINIDES:	**Ac** 89 2, 8, 18, 32, 18, 9, 2 Actinium	**Th** 90 232 2, 8, 18, 32, 18, 10, 2 Thorium	**Pa** 91 231 2, 8, 18, 32, 20, 9, 2 Protactinium	**U** 92 238 2, 8, 18, 32, 21, 9, 2 Uranium	**Np** 93 237 2, 8, 18, 32, 22, 9, 2 Neptunium	**Pu** 94 2, 8, 18, 32, 24, 8, 2 Plutonium

*Gases at standard temperature and pressure

			Col. 3	Col. 4	Col. 5	Col. 6	Col. 7	Col. 8
								He 2 4·00 2 0·179 Helium
			B 5 10·8 2, 3 2·34 × 10^3 Boron	**C** 6 12·0 2, 4 2·25 × 10^3 Carbon	**N** 7 14·0 2, 5 1·25 Nitrogen	**O** 8 16·0 2, 6 1·43 Oxygen	**F** 9 19·0 2, 7 1·70 Fluorine	**Ne** 10 20·2 2, 8 0·90 Neon
			Al 13 27·0 2, 8, 3 2·70 × 10^3 Aluminium	**Si** 14 28·1 2, 8, 4 2·33 × 10^3 Silicon	**P** 15 31·0 2, 8, 5 1·82 × 10^3 Phosphorus	**S** 16 32·1 2, 8, 6 2·07 × 10^3 Sulphur	**Cl** 17 35·5 2, 8, 7 3·21 Chlorine	**Ar** 18 39·9 2, 8, 8 1·78 Argon
Ni 28 58·7 2, 8, 16, 2 8·90 × 10^3 Nickel	**Cu** 29 63·5 2, 8, 18, 1 8·92 × 10^3 Copper	**Zn** 30 65·4 2, 8, 18, 2 7·14 × 10^3 Zinc	**Ga** 31 69·7 2, 8, 18, 3 5·90 × 10^3 Gallium	**Ge** 32 72·6 2, 8, 18, 4 5·35 × 10^3 Germanium	**As** 33 74·9 2, 8, 18, 5 5·73 × 10^3 Arsenic	**Se** 34 79·0 2, 8, 18, 6 4·81 × 10^3 Selenium	**Br** 35 80·0 2, 8, 18, 7 3·12 × 10^3 Bromine	**Kr** 36 83·8 2, 8, 18, 8 3·71 Krypton
Pd 46 106 2, 8, 18, 18, 0 12·0 × 10^3 Palladium	**Ag** 47 108 2, 8, 18, 18, 1 10·5 × 10^3 Silver	**Cd** 48 112 2, 8, 18, 18, 2 8·64 × 10^3 Cadmium	**In** 49 115 2, 8, 18, 18, 3 7·31 × 10^3 Indium	**Sn** 50 119 2, 8, 18, 18, 4 7·28 × 10^3 Tin	**Sb** 51 122 2, 8, 18, 18, 5 6·68 × 10^3 Antimony	**Te** 52 128 2, 8, 18, 18, 6 6·25 × 10^3 Tellurium	**I** 53 127 2, 8, 18, 18, 7 4·93 × 10^3 Iodine	**Xe** 54 131 2, 8, 18, 18, 8 5·85 Xenon
Pt 78 195 2, 8, 18, 32, 17, 1 21·5 × 10^3 Platinum	**Au** 79 197 2, 8, 18, 32, 18, 1 19·3 × 10^3 Gold	**Hg** 80 201 2, 8, 18, 32, 18, 2 13·6 × 10^3 Mercury	**Tl** 81 204 2, 8, 18, 32, 18, 3 11·8 × 10^3 Thallium	**Pb** 82 207 2, 8, 18, 32, 18, 4 11·3 × 10^3 Lead	**Bi** 83 209 2, 8, 18, 32, 18, 5 9·80 × 10^3 Bismuth	**Po** 84 2, 8, 18, 32, 18, 6 9·4 × 10^3 Polonium	**At** 85 2, 8, 18, 32, 18, 7 Astatine	**Rn** 86 2, 8, 18, 32, 18, 8 Radon

Eu 63 152 2, 8, 18, 25, 8, 2 Europium	**Gd** 64 157 2, 8, 18, 25, 9, 2 Gadolinium	**Tb** 65 159 2, 8, 18, 27, 8, 2 Terbium	**Dy** 66 162 2, 8, 18, 28, 8, 2 Dysprosium	**Ho** 67 165 2, 8, 18, 29, 8, 2 Holmium	**Er** 68 167 2, 8, 18, 30, 8, 2 Erbium	**Tm** 69 169 2, 8, 18, 31, 8, 2 Thulium	**Yb** 70 173 2, 8, 18, 32, 8, 2 Ytterbium	**Lu** 71 175 2, 8, 18, 32, 9, 2 Lutetium
Am 95 2, 8, 18, 32, 25, 8, 2 Americium	**Cm** 96 2, 8, 18, 32, 25, 9, 2 Curium	**Bk** 97 2, 8, 18, 32, 26, 9, 2 Berkelium	**Cf** 98 2, 8, 18, 32, 28, 8, 2 Californium	**Es** 99 2, 8, 18, 32, 29, 8, 2 Einsteinium	**Fm** 100 2, 8, 18, 32, 30, 8, 2 Fermium	**Md** 101 2, 8, 18, 32, 31, 8, 2 Mendelevium	**No** 102 2, 8, 18, 32, 32, 8, 2 Nobelium	**Lr** 103 2, 8, 18, 32, 32, 9, 2 Lawrencium

Properties of the Elements

Key:

Symbol	Atomic Number
	m.p. (°C)
	b.p. (°C)
1st Ionisation Energies ($J\ mol^{-1} \times 10^3$)	Electro-negativity (Pauling)
2nd	Covalent Radius ($m \times 10^{-10}$)
3rd	(Charge on Ion)*
4th	Ionic Radius ($m \times 10^{-10}$)

Symbol	Atomic Number	m.p. (°C)	b.p. (°C)	1st	2nd	3rd	4th	Electro-negativity	Covalent Radius	Charge on Ion	Ionic Radius
H	1	−259	−253	1310				2·1	0·37	(1−)	2·08
Li	3	179	1317	520	7280	11800		1·0	1·23	(1+)	0·68
Be	4	1280	2970	900	1760	14900		1·5	0·89	(2+)	0·35
Na	11	98	892	490	4560			0·9	1·57	(1+)	1·10
Mg	12	651	1110	740	1450	7740		1·2	1·36	(2+)	0·80
K	19	64	774	420	3070			0·8	2·03	(1+)	1·46
Ca	20	843	1490	590	1150	4940		1·0	1·74	(2+)	1·08
Sc	21	1540	2730	630	1240	2390	7120	1·3	1·44	(3+)	0·81
Ti	22	1680	3260	660	1310	2720	4170	1·5	1·32	(2+)	0·94
V	23	1890	3000	650	1370	2870	4600	1·6	1·22	(2+)	0·87
Cr	24	1890	2480	650	1590	2990	4770	1·6	1·17	(2+)	0·90
Mn	25	1240	2100	720	1510	3250	5190	1·5	1·17	(2+)	0·90
Fe	26	1540	3000	760	1560	2960	5400	1·8	1·16	(2+)	0·85
Co	27	1500	2900	760	1650	3230	5100	1·8	1·16	(2+)	0·82
Rb	37	39	688	400	2650			0·8	2·16	(1+)	1·57
Sr	38	769	1380	550	1060			1·0	1·91	(2+)	1·24
Y	39	1500	2930	640	1180			1·2	1·62	(3+)	0·97
Zr	40	2980		670	1270			1·4	1·45		
Nb	41	2470	4930	650	1380			1·6	1·34		
Mo	42	2610	5560	690	1560			1·8	1·29		
Tc	43	2200	3500	700	1470			1·9			
Ru	44	2250	3900	720	1620			2·2	1·24		
Rh	45	1970	3730	750	1750			2·2	1·25		
Cs	55	29	690	380	2420			0·7	2·35	(1+)	1·78
Ba	56	725	1140	500	970			0·9	1·98	(2+)	1·44
La	57	920	3470	540	1100			1·1	1·69	(3+)	1·14
Hf	72	2150	5400	530	1440			1·3			
Ta	73	2300	5430	580	1570			1·5	1·44		
W	74	3410	5930	770	1710			1·7	1·30		
Re	75	3180	5630	760	1600			1·9	1·28		
Os	76	3000	5000	840	1630			2·2	1·26		
Ir	77	2410	4530	890				2·2	1·26		
Fr	87							0·7			
Ra	88	700	1750	510	980			0·9			
Ac	89	3200		670	1170			1·1			

*The Ionic Charge quoted for an element in this table should not be taken to mean that the element always exists as that ion in its compounds, e.g.

hydrogen does not usually form H^-,

carbon does not usually form C^{4+}.

								He 2 −272 −269 2370 5250
			B 5 2300 2550 800 2·0 2430 0·80 3660 (3+) 25000 0·20	**C** 6 3650 4200 1090 2·5 2350 0·77 4600 (4+) 6240 0·15	**N** 7 −210 −196 1400 3·0 2860 0·74 4600 (3−) 7480 1·71	**O** 8 −218 −183 1310 3·5 3390 0·74 5320 (2−) 7460 1·30	**F** 9 −220 −188 1680 4·0 3370 0·72 6030 (1−) 8410 1·23	**Ne** 10 −249 −246 2080 3960 6200 9380
			Al 13 660 2470 580 1·5 1820 1·25 2750 (3+) 11600 0·61	**Si** 14 1410 2360 790 1·8 1580 1·17 3230 (4+) 0·48	**P** 15 44 280 1060 2·1 1900 1·10 2920 (3−) 2·12	**S** 16 113 445 970 2·5 2260 1·04 3390 (2−) 1·72	**Cl** 17 −101 −35 1260 3·0 2300 0·99 3850 (1−) 1·72	**Ar** 18 −189 −186 1520
Ni 28 1450 2730 740 1·8 1750 1·15 3390 (2+) 5400 0·78	**Cu** 29 1080 2600 750 1·9 1960 1·17 3550 (2+) 5700 0·81	**Zn** 30 419 907 910 1·6 1730 1·25 3830 (2+) 5990 0·83	**Ga** 31 30 2400 580 1·6 1980 1·25 2960 (3+) 6200 0·70	**Ge** 32 937 2830 760 1·8 1540 1·22 3300 (4+) 0·62	**As** 33 817 613 970 2·0 1950 1·21 2730 (3−) 2·22	**Se** 34 217 685 940 2·4 2080 1·17 3090 (2−) 1·88	**Br** 35 −7 59 1140 2·8 2080 1·14 3470 (1−) 1·88	**Kr** 36 −157 −152 1350
Pd 46 1550 2930 800 2·2 1880 1·28 (2+) 0·94	**Ag** 47 960 2210 730 1·9 2070 1·34 (1+) 1·23	**Cd** 48 321 765 870 1·7 1630 1·41 (2+) 1·03	**In** 49 157 2000 560 1·7 1820 1·50 (3+) 0·87	**Sn** 50 232 2260 710 1·8 1410 1·40 2940 (2+) 3930 1·12	**Sb** 51 630 1380 830 1·9 1590 1·41 (3−) 2·45	**Te** 52 450 990 870 2·1 1800 1·37 (2−) 2·21	**I** 53 114 184 1010 2·5 1840 1·33 (1−) 2·13	**Xe** 54 −112 −107 1170
Pt 78 1770 3900 870 2·2 1870 1·29	**Au** 79 1060 2970 890 2·4 1980 1·34	**Hg** 80 −39 357 1010 1·9 1810 1·44 (2+) 1·10	**Tl** 81 304 1460 590 1·8 1970 1·55 (3+) 0·96	**Pb** 82 328 1740 720 1·8 1450 1·54 (2+) 1·26	**Bi** 83 271 1560 770 1·9 1610 1·52	**Po** 84 254 962 810 2·0	**At** 85 2·2 1·40	**Rn** 86 −71 −62 1040

Approximate Atomic Weights—for Calculations

Element	Symbol	Atomic Weight
Aluminium	Al	27
Argon	Ar	40
Barium	Ba	137
Beryllium	Be	9
Boron	B	11
Bromine	Br	80
Calcium	Ca	40
Carbon	C	12
Chlorine	Cl	35·5
Chromium	Cr	52
Cobalt	Co	59
Copper	Cu	63·5
Fluorine	F	19
Helium	He	4
Hydrogen	H	1
Iodine	I	127
Iron	Fe	56
Krypton	Kr	84
Lead	Pb	207
Lithium	Li	7
Magnesium	Mg	24
Manganese	Mn	55
Mercury	Hg	201
Neon	Ne	20
Nickel	Ni	58·7
Nitrogen	N	14
Oxygen	O	16
Phosphorus	P	31
Potassium	K	39
Silicon	Si	28
Silver	Ag	108
Sodium	Na	23
Sulphur	S	32
Tin	Sn	119
Xenon	Xe	131
Zinc	Zn	65·4

Common Ions and their Charges

Ion	Formula	Ion	Formula
Ammonium	NH_4^+	Nitrate	NO_3^-
Carbonate	CO_3^{2-}	Nitrite	NO_2^-
Hydrogen carbonate	HCO_3^-	Phosphate	PO_4^{3-}
Hydrogen sulphate	HSO_4^-	Sulphate	SO_4^{2-}
Hydrogen sulphite	HSO_3^-	Sulphite	SO_3^{2-}
Hydroxide	OH^-		

Standard Reduction Electrode Potentials (I.U.P.A.C. convention)

$Li^{+}(aq) + e$	$\rightarrow Li(s)$	−3·02 V
$Cs^{+}(aq) + e$	$\rightarrow Cs(s)$	−2·92
$Rb^{+}(aq) + e$	$\rightarrow Rb(s)$	−2·92
$K^{+}(aq) + e$	$\rightarrow K(s)$	−2·92
$Ca^{2+}(aq) + 2e$	$\rightarrow Ca(s)$	−2·87
$Na^{+}(aq) + e$	$\rightarrow Na(s)$	−2·71
$Mg^{2+}(aq) + 2e$	$\rightarrow Mg(s)$	−2·37
$Al^{3+}(aq) + 3e$	$\rightarrow Al(s)$	−1·66
$Zn^{2+}(aq) + 2e$	$\rightarrow Zn(s)$	−0·76
$Cr^{3+}(aq) + 3e$	$\rightarrow Cr(s)$	−0·74
$S(s) + 2e$	$\rightarrow S^{2-}(aq)$	−0·51
$Fe^{2+}(aq) + 2e$	$\rightarrow Fe(s)$	−0·44
$Cr^{3+}(aq) + e$	$\rightarrow Cr^{2+}(aq)$	−0·41
$Sn^{2+}(aq) + 2e$	$\rightarrow Sn(s)$	−0·14
$Pb^{2+}(aq) + 2e$	$\rightarrow Pb(s)$	−0·13
$Fe^{3+}(aq) + 3e$	$\rightarrow Fe(s)$	−0·04
$2H^{+}(aq) + 2e$	$\rightarrow H_2(g)$	0·00
$S(s) + 2H^{+}(aq) + 2e$	$\rightarrow H_2S(aq)$	0·14
$Sn^{4+}(aq) + 2e$	$\rightarrow Sn^{2+}(aq)$	0·15
$Cu^{2+}(aq) + e$	$\rightarrow Cu^{+}(aq)$	0·15
$SO_4^{2-}(aq) + 2H^{+}(aq) + 2e$	$\rightarrow SO_3^{2-}(aq) + H_2O$	0·17
$Hg_2Cl_2(s) + 2e$	$\rightarrow 2Hg(l) + 2Cl^{-}(aq)$	0·27
$Cu^{2+}(aq) + 2e$	$\rightarrow Cu(s)$	0·34
$Cu^{+}(aq) + e$	$\rightarrow Cu(s)$	0·52
$I_2(s) + 2e$	$\rightarrow 2I^{-}(aq)$	0·54
$Fe^{3+}(aq) + e$	$\rightarrow Fe^{2+}(aq)$	0·77
$Ag^{+}(aq) + e$	$\rightarrow Ag(s)$	0·80
$2NO_3^{-}(aq) + 4H^{+}(aq) + 2e$	$\rightarrow N_2O_4(g) + 2H_2O$	0·81
$Hg^{2+}(aq) + 2e$	$\rightarrow Hg(l)$	0·85
$NO_3^{-}(aq) + 4H^{+}(aq) + 3e$	$\rightarrow NO(g) + 2H_2O$	0·96
$Br_2(l) + 2e$	$\rightarrow 2Br^{-}(aq)$	1·07
$O_2(g) + 4H^{+}(aq) + 4e$	$\rightarrow 2H_2O$	1·23
$MnO_2(s) + 4H^{+}(aq) + 2e$	$\rightarrow Mn^{2+}(aq) + 2H_2O$	1·23
$Cr_2O_7^{2-}(aq) + 14H^{+}(aq) + 6e$	$\rightarrow 2Cr^{3+}(aq) + 7H_2O$	1·33
$Cl_2(aq) + 2e$	$\rightarrow 2Cl^{-}(aq)$	1·36
$Au^{3+}(aq) + 3e$	$\rightarrow Au(s)$	1·50
$MnO_4^{-}(aq) + 8H^{+}(aq) + 5e$	$\rightarrow Mn^{2+}(aq) + 4H_2O$	1·51
$Au^{+}(aq) + e$	$\rightarrow Au(s)$	1·68
$F_2(g) + 2e$	$\rightarrow 2F^{-}(aq)$	2·87

Trivial Names of some common Organic Compounds

Systematic Name	Trivial Name
Ethene	Ethylene
Ethyne	Acetylene
Chloroethene	Vinyl chloride
Trichloromethane	Chloroform
Tetrachloromethane	Carbon tetrachloride
Methanol	Methyl alcohol
Ethanol	Ethyl alcohol
Propan-2-ol	Isopropyl alcohol
Butan-2-ol	*s*-Butyl alcohol
2-methylpropan-2-ol	*t*-Butyl alcohol
Methanal	Formaldehyde
Ethanal	Acetaldehyde
Propanone	Acetone
Methanoic acid	Formic acid
Ethanoic acid	Acetic acid
Propanoic acid	Propionic acid
Methanoate	Formate
Ethanoate	Acetate
Methanoyl	Formyl
Ethanoyl	Acetyl

Index